AF328775

KAY FIRTH-BUTTERFIELD

COEXISTING WITH AI

WORK, *LOVE*, AND *PLAY* IN A CHANGING WORLD

WILEY

Library of Congress Cataloging-in-Publication Data is Available:

ISBN: 9781394278107 (cloth)
ISBN: 9781394278169 (ePub)
ISBN: 9781394278183 (ePDF)

Cover Design: Wiley
Cover Image: © amplion/stock.adobe.com, © nadia_bormotova/Getty Images

To those who have believed in me over the years. There are too many to name; you know who you are. The people named here recognized that the things I had to say about AI should be heard and elevated my voice. Thank you.

Walter Burrough
Rohaise Firth-Butterfield
John Havens
Tom Meredith
Brad Smith
Kai-Fu Lee
Will-i-am
Arvind Krishna
Vilas Dhar
Murat Sonmez
Jeremy Jurgens
Michael Stewart
Victoria Savanh

Contents

CHAPTER 1

The Ultimate Question of Life, the Universe, and Everything

All the world's a stage,
*   And all the men and women merely players;*
*   They have their exits and their entrances,*
*   And one man in his time plays many parts,*
*   His acts being seven ages.*
                    —William Shakespeare, *As You Like It*

The purpose of this book is to offer you a great deal of information about how artificial intelligence (AI) affects our present and will shape our future. How you decide to use AI and what you think about it is your decision; I don't seek to make up your mind for you. The objective is simply to empower you to co-exist with AI in the ways you choose.

For many, the pace at which we are seeing AI enter our lives and the lives of our loved ones is overwhelming. We don't know what it is, how it works, or where it's going, but we do know that it's everywhere. Are we now, or will we ever be, as Shakespeare puts it, mere players dancing on a stage while the AI works for us, understands us, or manipulates us? Perhaps we will live lives of untold fulfillment and riches because AI manages and optimizes everything for us. To answer these questions, it's time for us all to learn more and make the right decisions now about the part we humans will play in the age of AI.

This book is your guide to the most significant—and potentially dangerous—invention of our time, indeed, ever for humans. Welcome to the future! Read on to join the discussion of what we humans want from AI and what we don't. This is an essential guide to helping you have your say about where AI should go and to enabling you to participate in essential conversations at work and in your personal life.

The Questions of AI

Let's start by posing some questions. Proponents of AI say it will make our lives as humans, and our planet, better. For example, we will not have to do as many tedious jobs and we will be freed to do the more interesting and intellectual jobs. In a July 2025 report, Microsoft said that AI will be doing the jobs of everything from historians to telemarketers to models and writers giving them all time to do what they enjoy rather than the drudgery of work.[1] As current generative AI models have an up to 60% likelihood of giving the wrong answer, this may cause more problems than are solved (see "Hallucinations").

If this is really happening, does it make sense that we are running to catch up with AI as opposed to AI being implemented, after careful thought, at a speed that humans can manage and adapt to? Also, is it being designed, developed, and used purposefully with our best interests at heart? By the time you finish the book, you will be well equipped to answer these questions.

There is a lot of talk about AI in the halls of power in countries around the world, in businesses, in seats of learning, and everywhere in the media. AI is said to be on track to add a cumulative $19.9 trillion by 2030 or 3.5% to the world's economy.[2] However, that is based on the spending and income of the Big Tech companies that produce AI, largely based in the United States and China. In early January 2025 we saw that assumption as a marker of economic growth poleaxed with Nvidia, the world's largest maker of graphics processing units (GPUs) (the chips needed to create AI systems) losing $600 billion of market valuation, but it bounced back within a few months benefiting from deals with many Middle Eastern companies. Goldman

Sachs predicted a 7% growth in global gross domestic product (GDP) due to AI by 2034.[3]

In January 2024 the International Monetary Fund noted on their blog that not only would 40% of jobs be affected by AI but also "historically, automation and information technology have tended to affect routine tasks, but one of the things that sets AI apart is its ability to impact high-skilled jobs. As a result, advanced economies face greater risks from AI—but also more opportunities to leverage its benefits—compared with emerging market and developing economies."[4]

If you see human well-being in terms of money, AI should bring huge economic benefits, although there is likely to be more benefit to those in the countries with robust use of AI further developing the global and national divides between those who have and those who do not.

If you see human well-being as having time to do whatever you want because goods are cheap or free and resources are unlimited, then AI can also lead to that life. It's called the *abundance economy* and is best portrayed by *Star Trek*, in which everything is provided for free, leaving humans to do what they wish, in that case exploring the universe.[5] Other examples of this can be found elsewhere in sci-fi such as in E. M. Forster's *The Machine Stops*,[6] *Singularity Sky* by Charles Stross,[7] and *WALL-E*.[8]

Humans have always dreamed of liberation from work; it was even a core tenant of communism. Should we reach abundance before we reach space, then we need to figure out, and as soon as possible, what billions of people will do with unlimited free time. History has shown that humans with time on their hands need ways of being entertained, so the joint developments of gaming, metaverse, social media, and more might have developed to provide constant entertainment. It's easy to see AI companies like Replika[9] and Character.ai,[10] which make chatbot "friends" for humans, fulfilling the constant attention that humans will need. Already, it is known that many Americans use a chatbot for an intimate partner.[11] The benefits of such AI "friends" have been explored by Voice of America.[12] They worked with three people who had chosen relationships with AI rather than with humans. One of them had been talking to the AI bot for four months but actually had the equivalent of three years of interactions because he was "talking" to "her" so much. They conclude that 20% of Americans have used a personal chatbot of this type. Given that many young people are eschewing deeper relationships with humans, this could be one of the new ways in which humans create happiness for themselves.[13] Additionally, the jury is out as to whether the abundance economy will be available to everyone on the planet or whether it too will further divide us. Privacy is an ever-present "push you/pull me" when using AI. For example, in May 2025, Replika was fined $5.64 million by the Italian privacy regulators for "not having the legal basis for processing users' data and had no age-verification system to restrict children from accessing the service. . . ."[14]

For those of you who see human well-being in work, paid or charitable, then you may not want AI to be doing your job. However, there are plenty of things that humans can do, which, currently, neither AI nor AI-enabled robots can do.

If you have read this far, it is probably because you want to make decisions about how you want to use and interact with AI to benefit your life now, in the future, and for your children and beyond. You want to be able shape your life, understanding and mitigating how AI will affect your hopes, fears, and dreams.

Stages of Life with AI

In his play *As You Like It*,[15] Shakespeare, arguably one of the finest human storytellers, used the character Jaques to describe human lives as having seven stages:

- The Infant
- The Schoolboy
- The Lover
- The Soldier
- The Justice
- The Pantaloon
- Old Age

(Of course, he was writing in the 1500s and 1600s and so gender was assumed male in this piece. That assumption is not made in this book.)

As you turn the pages of this book, you will find a wealth of ways in which AI is affecting your life, country, the planet, and children today and tomorrow. The book is built on these seven stages with information on health care, geopolitics, education, work, business transformation, and much more. However, you will soon see that, given the ubiquitous existence of AI, any conversation in one stage could probably fit into any of the other stages. For example, anyone of any age or gender might be interested in the use of AI in warfare, so you'll find that information in Chapter 5.

In May 2024 the BBC, Oxford University, and Reuters published a study into general knowledge of AI tools.[16] Whether you are one of the 53% of Americans who know nothing about an AI called *ChatGPT* (there are other figures for other parts of the world where computers are used but a split about the 50% mark is common) or someone who uses it regularly in the course of work and/or school, there is something new for you to discover.

Wherever you are on the continuum, you can use this book as a guide to where we already find AI in our lives, where we may find AI in the future, and what protections we might want to put around use of AI to benefit our children and descendants.

Shifting Global Order and AI

AI comes at a time when we humans are already trying to cope with a number of other problems and novel changes. To the complexities of AI we can add the following issues, which are capable of causing an imbalance, for good or ill, in our lives:

- **Moving into a new world order.** Our most recent global elections look set to effect the greatest change in the geopolitical landscape since the end of World War II. No one seems to know the extent to which the use of AI to create misinformation and disinformation during the election cycles led to this change, but it was undoubtedly a factor. There are many books written about this phenomenon so it is not covered here. However, the ability of bad actors to use AI in our social media feeds should never be underestimated. It is becoming increasingly harder for good actors to unmask the bad ones, and with the loosening of content control by social media companies this problem can only get much worse. In cybersecurity operations it's possible to use AI to find AI-created scams and untruths, resulting in an ever spiraling AI fight to win a position that is dominated by AI, whether in creation of material, defense of the truth, or spreading of rumor and untruths. One AI is looking for another bad acting AI, and humans don't have a role to play in this fast-moving technological battle.

◆ **Climate change.** Whether you believe it is created by humans or not, the climate is changing, bringing catastrophic weather events that dispossess millions of people around the world from their homes and livelihoods. It is expected that 143 million people in the Global South will migrate internally or across boundaries by 2050 driven out of their homes by their changing climate.[17] The economic consequences are enormous today and will continue to be so into the future. Where businesses are destroyed, the people affected not only lose their livelihood and their premises but also may have to move as well, leaving the area devoid of economic activity. This will, in turn, affect the ability of business owners to use AI to supercharge their businesses. They will need to ensure their data is secured off-site and maybe off country, machine-readable, and readily transportable. Immigrants forced by such circumstance into a new country rarely obtain a job at a level commensurate with the one they had at home. This reduces the economic well-being of the whole family. For example, an immigrant who is a doctor will need to find the money to retake medical exams in the new country. Where schools are destroyed and children cannot be educated, that area loses its ability to create a new generation of skilled, educated citizens. In this circumstance there may be a role for AI as teacher because, provided there is power, AI can educate many at the same time and at a speed that suits the learning level of the child. In turn, these displacements and changes affect the economy, and in many places these circumstances will increase the poverty levels of the area. In Bangladesh in 2024, six million people were displaced due to flooding[18] while in the United States in 2024 "61,685 fires burned 8,851,142 acres."[19]

AI can help mitigate weather events, making predictions that help people move to safety or to educate displaced children wherever they are, but AI only works if you have access to the internet, to power, and to a computer. Of the world's population, 2.6 billion people, or 33%, do not have internet access and so AI's positive contributions are not equitably spread around the world to provide help in such times of crisis.[20]

♦ **Birthrates falling, bringing new economic challenges.** Falling birthrates across the whole world put pressure on the stability of countries and cause their citizens to worry both about who will support them in their old age and whether their country's economy will fail. Arguments about immigration lead to further instability, but here AI has a real chance of helping. If AI can do jobs previously done by immigrants or new members of the workforce, it can make up the shortfall of human workers. Japan, which has a long-standing policy of limiting immigration, has been working hard on developing AI and robotics to enhance their workforce. AI doesn't even need to do the whole job; it could just help the human. Likewise, it's possible in the future that brain–computer interfaces with AI[21] can help solve dementia, with AI making exoskeletons even better to keep older citizens well and, should they choose it, working for longer.[22] Caring for older citizens with disabilities is extremely expensive for societies where families do not live together. Therefore, creating robots with embedded AI to make them smart can make those tasks, such as cooking, housework, mobility, and more, which humans get less able to do as they grow old easier to do.[23]

♦ **Aging workforce.** In the Global North many workers are growing older. A great example of this is skilled construction workers in the United States, who have an average age of 42[24]; there are also fewer young people choosing this career. Using robots, like Boston Dynamic's Spot the dog, can enable workers to be safer and have some of their workloads be carried by a machine.[25] Once humanoid robots enabled with AI are more widely available they can do jobs that humans don't want or are unsafe for them. Likewise, as we have seen, older humans are able to use AI reliably and the AI itself is reliable they will be able to work longer, should they choose to do so.

- **Falling social cohesion.** In the Global North we are becoming, according to Derek Thompson, increasingly antisocial. He says, "Americans are now spending more time alone than ever. It's changing our personalities, our politics, and even our relationship to reality." In support, he shares this staggering statistic: "In 2023, 74% of all restaurant traffic came from 'off premises' customers—that is, from takeout and delivery—up from 61% before COVID, according to the National Restaurant Association."[26] Thompson suggests that this is just one form of aloneness that is affecting our ways of communicating with each other. We have seen many people turn to AI rather than people for companionship, affection, validation, and friendship. For example, many young people have a string of friendships with other gamers whom they met through the online games they play and an AI friend. This changes the way in which people communicate, often in text not verbally.

- **Changing notions of privacy.** Anyone who has lived in a small community knows that it's hard to keep secrets from others. Over time, as many moved into cities we became more individualized and developed a love for keeping some aspects of our lives private. This has become the default position for most peoples, especially in the Global North. Although there had been treaties and philosophical discussion of human rights from ancient times the modern conversation started in 1948 with the Universal Declaration of Human Rights. After World War II many wanted to enshrine our rights as human beings not to have everyone know everything about us, particularly not governments and companies. One of those rights is a right to privacy.[27] Social media has changed our notions of privacy because we voluntarily offer details about our lives to the world, and then later using privacy settings to constrain who can see that information. AI brings new threats to our privacy because, as we will see, it needs massive amounts of data about us and the world in order to be able to function well. In answer to some of these challenges governments have created legislation to protect the privacy of humans. In Europe they passed the General Data Protection Regulation (GDPR) in 2018. It gives powers to act against companies that impugn the privacy of its citizens worldwide. Many other countries around the world enacted their own versions of the GDPR and the new European Union (EU) AI Act attaches the tenants of the GDPR to AI systems.

♦ **AI, the perfect surveillance tool.** Without data AI cannot exist. This means that all the companies and governments that have made it their objective to collect lots of data from their customers and citizens are in the unique position of being able to know a great deal about them. Indeed, you may have willingly given up this data, for example, to a social media company, or have been compelled to do so, for example, when filing your tax returns. The more that AI can learn about you from data the better it is able to surveil you.[28] This can be a huge benefit because it learns what you like and then offers you things to buy that you will like or it can be a burden if used as a political weapon by a dictatorship. Of course, personalization of products may also stop you expanding your style horizons, leading to everyone choosing the same type and color of furniture or clothes.

♦ **Turning to AI as an omnipotent power.** AI is being sold to us as omnipotent and able to cure all the ills of our lives and planet. Eric Schmit, former CEO of Google, for example, tells us that "advancing AI will eventually solve the climate crisis"[29] while also, in April 2025, telling the US Congress that AI would consume 99% of the world's energy output by 2030. AI shares some apparent traits with a religious God in that it knows a great deal about you, probably more than you know yourself, and, in the future, it may decide your fate, but we must remember that it is simply a machine built by clever people to an idea first conceived by sci-fi writers. We are also told that AI will cure loneliness, mental health problems, give personalized education to all, cure cancer, and more. While it may help in some or all of these situations, this does not make it an omnipotent being. *MIT Press Reader* suggested in 2024 that Silicon Valley had a tendency to view AI as God-like.[30] Anthony Levandowski created his own brand of AI worshiping faith in the mid-2010s and closed it in 2020. But he is a visible side of the mania in Silicon Valley about AI which, writer Greg Epstein says has layers of religiosity to it. For example "beliefs and rituals that resemble religion—complete with digital deities, moral codes, and threats of damnation."[31]

Many people seem to think that AI is a magic wand, or perhaps fairy dust. Sprinkle a problem with AI and it will be solved. We are not there—yet—and may never get there because of the type of AI we are currently prioritizing. There are many scientists who believe that we cannot achieve superintelligent AI (defined in the next section) even if possible through creating ever better large language models (LLMs) (defined next).[32]

WHAT IS A LARGE LANGUAGE MODEL?

By noting which words are used together with other words, LLMs construct a "model" of understanding. The biggest and most able LLMs are generative pre-trained transformers known as GPTs, which you will mainly come across as chatbots such as ChatGPT, Claude, Mistral, Grok, DeepSeek, LLAMA, and Gemini. You will notice that only one mentions GPT in its name but they are all using those generative pre-trained transformers. For specific tasks LLMs can be fine-tuned. This happens when the model are changed, usually by supervised learning, which is done by humans. The chatbots can also be guided to their answers by asking questions, known as prompting. They appear to be extremely intelligent and can perform many tasks as well or better than humans, but they carry inherent problems that limit their reliability, which we will look at in the chapters to come. There are many books on generative AI if you want to do a deep dive but more I will provide some definitions in this book.

There is debate about the abilities of these models to do what they are hyped to be able to do and to be able to achieve superintelligence in the next few years, so it may be worth a wry smile that even in science fiction, *Star Trek*'s superintelligent AI Zora didn't appear until 2257.[33] In the same way the enormous supercomputer featured in *The Hitch Hikers Guide* to the Galaxy, named Deep Thought, came up with the answer to the Ultimate Question of Life, the Universe, and Everything but it took 7.5 million years.[34] If AI is to help our planet in the way some think it can, then we need to have it become even more useful very soon. Time is running out for our planet and that is exacerbated by the huge need for energy and water created by AI systems.[35] Are we in a position to understand and benefit from superintelligent AI systems if they appear in 2026, or will the benefits be specifically for a few? There have been many movies about this but perhaps *Elysium* is the best known? You may also be interested in watching the TV series Black Mirror, which asks some of the same questions you will discover in this book and provides one, not the only, answer.

Yet another question for us to answer is that if we surround ourselves with intelligent computers, do we become less intelligent? If so, then the scenario that computers run the world for us may be inevitable. Studies show that our addiction to social media has led to our having a decreased attention span.[36] Equally, social media may be leading to declining rates of children in the United States and United Kingdom being able to read.[37] If this continues, will we even be able to concentrate sufficiently to run our world? Like pilots, will we have to keep training over and over again to ensure that we humans stay in control? A study in May 2025 showed that students were using a chatbot so often to complete homework assignments that when they graduated they had not been educated to think critically or sufficiently in their subject matter topic.[38] Then another study published by MIT LAB in June 2025 showed that

> LLM users also struggled to accurately quote their own work. While LLMs offer immediate convenience, our findings highlight potential cognitive costs. Over four months, LLM users consistently underperformed at neural, linguistic, and behavioral levels. These results raise concerns about the long-term educational implications of LLM reliance and underscore the need for deeper inquiry into AI's role in learning.[39]

If we are to thrive as humans in the age of AI we need to be highly trained in critical thinking because developers of AI can train their AI to give a particular narrative even when not asked about the issue! In May 2025, Grok, an AI owned by Elon Musk, started talking about "white genocide" in South Africa even though it had not been asked about it. "White genocide" is a highly debated issue that Musk believes in. The gratuitous information came at a time when white South Africans arrived in the United States as refugees. Whether or not Musk had anything to do with this random behavior of Grok, it shows that LLMs can be easily manipulated by users and owners. Indeed, it doesn't matter which political beliefs you hold. LLMs can further polarize the world's populations.[40]

Everything You Always Wanted to Know About AI but Were Afraid to Ask

Since human's earliest tales, we have been obsessed with escaping our natural sphere. We are not built to fly, but Icarus designed wings that enabled him to do so. His and other "escapes" also come with a cautionary tale: he flew too close to the sun and lost his life. Just as we have perennial tales of good and evil so do we have tales of those who go beyond the human form or mind and are punished for it or just end up hurt or dead, depending on your belief system. And yet it is intensely human to strive for more and those who do are lauded, often with money as well as praise. Without these explorers, humans would not have moved from walking on all fours. So where does AI fit into the narrative—are we about to fly too close to the sun?

THE HISTORY OF AUTOMATION

Automata and thinking machines have long historical roots. The Greeks built steam-powered doves. Leonardo da Vinci built a "mechanical knight" that was able to sit down, stand up, move its head, and lift its visor.[41] Jaquet-Droz built a clockwork boy in 1774 that could draw.[42] In 1818, Mary Shelley wrote what is widely considered the first substantive work of science fiction in which she described the cautionary tale of Frankenstein, a thinking machine of reanimated body parts. She was a contemporary of the first "computer programmer" who coincidentally was also part of the inner circle of English Romantic poets. Mary Shelley was married to Percy Bysshe Shelley and Ada Lovelace was the daughter of Lord Byron.

Ada Lovelace was a fine mathematician. Because of her social standing and somewhat looser attitude about Victorian rules, she was able to work with Charles Babbage, a famous and brilliant mathematician. While he invented the "analytical engine" it was she who wrote the notes, and possibly, Note G, which included the first published algorithm (although that's the subject of controversy). The analytical engine was not produced during either her or Babbage's lifetime but has the same structure that eventually became the electronic computer. Lovelace wrote that "the Analytical Engine has no pretensions whatever to originate anything. It can do whatever we know how to order it to perform. It can follow analysis; but it has no power of anticipating any analytical relations or truths."[43] That can equally be said of our current AI-enabled devices.

In his play written in 1920, *Rossum's Universal Robots*, Karel Čapek first used the word *robot*. He described a being that was created by humans and could be used to do tasks that humans didn't want to do or were unable to do. AI, as we know it, was first described in science in 1950 by Alan Turing, famous for inventing the Enigma Machine during World War II. He wrote a paper called "Computing Machinery and Intelligence" in which he mused about whether a machine could "think."[44] Because of the definition of the word *think* he devised a test that we now call the *Turing test*, which is one of the tests we use to judge whether an AI is as capable as a human (more usually known as artificial general intelligence [AGI] or superintelligence). Then in 1966 John McCarthy, a mathematician at the time, organized a summer camp at Dartmouth, USA, "to proceed on the basis of the conjecture that every aspect of learning or any other feature of intelligence can in principle be so precisely described that a machine can be made to simulate it." It was at that conference that the term *artificial intelligence* was first used.[45]

At that time, it was generally thought that machines would need to have the world explained to them in code to be able to show signs of intelligence. In other words, for a computer to understand a cat it would need each characteristic of a cat explained to it. However, humans identify cats by pattern matching—"I have seen a cat and this is the same as the one I have seen so it too must be a cat." In fact, our abilities go well beyond that because we recognize cartoon cats as cats and lions as cats, too. It was realized that if a computer was exposed to enough data (pictures of cats) then it could be trained to identify other pictures of cats. This is known as *computer vision.*

Deep learning, the term coined by Geoffry Hinton and developed by him and other researchers, was a type of AI development that used layers of neural nets to work across lots of data that enabled the machine to recognize cats and other images without having been trained to understand all the characteristics of a cat or whatever it had been shown. In 2016, DeepMind a London-based AI company (now owned by Google) showed that this way of training an AI was very powerful. Its computer, AlphaGo, beat the human Go world champion, Lee Sidol. How did that happen?

The AI model was trained (developed knowledge of all the moves and strategies in Go) using machine learning and then used a "Monte Carlo tree search algorithm" to find its moves. In this case, the machine learning was what is called a *deep learning method using a neural network*. The term *neural network* is confusing because it sounds as it is related to the neurons in the brain. That's not the case. In this event the neural network in AI took data from humans and other machines that had played the game and was trained to identify the moves that are most efficient and most likely to win. Every time the machine plays the neural network strengthens the search so the machine gets better and better the more that it plays. Interestingly it also came up with an entirely new move that clinched its victory. In Go we are now in the position that no human can beat the machine.

It is the success of this machine learning method (deep learning) behind the LLMs that power the generative AI we have now. Although OpenAI had been working on generative AI since 2016, the first the world at large heard about it was in November 2022 when OpenAI released version 3.5 of a model called *ChatGPT*.

Since then, there have been more generative AI models produced by companies in the United States, Europe, and China. Currently, none are significantly better than others but the battle is on. The models developed in the United States include ChatGPT (OpenAI), Claude (Anthropic), Grok (xAI), Gemini (Google), and Llama (Meta). In Europe, there are Mistral and Aleph Alpha, and in China the model known as DeepSeek (V-3) appears to be performing at about the level of the current offering from OpenAI. There is a great deal of interest in DeepSeek because it is reported that it costs only $6 million to train, which is a miniscule fraction of the money spent by the Western companies in training their models of equal ability. There are now some suggestions that it was actually built on a Western model, possibly the open-source Meta tool. Other companies with generative AI models in China include

Baidu (Ernie), Alibaba, Tencent (Hunyuan and Yuanbao chatbot), and Zhipu AI. In the United Arab Emirates a company called G42 has developed an Arabic language LLM. Former President Biden's order regarding the sale of US GPUs (chips) might have had a significant impact on further development of AI in China and elsewhere in the world. However, during a state visit to the Middle East President Trump relaxed the selling of chips to allies there. There are many other companies around the world developing generative AI but currently the ones mentioned in this section hold the commercial advantage. In 2025, the Swiss created a sovereign LLM that is claimed to be "as neutral as Switzerland."[46]

LET'S DEFINE AI

Let's start with an easy definition. There is a legal definition agreed by 47 countries (including the United States in 2019 and all of the EU nations):

> An AI system is a machine-based system that, for explicit or implicit objectives, infers, from the input it receives, how to generate outputs such as predictions, content, recommendations, or decisions that can influence physical or virtual environments.[47]

The EU describes generative AI as

> the use of AI to create new content, like text, images, music, audio, and videos.

It also provides an extra definition of how the models are created:

> Generative AI models are trained on very large datasets from which they learn the patterns and structure and then generate new synthetic content that has similar characteristics.[48]

This definition has recently come under pressure from the Chinese generative AI system, DeepSeek, which used considerably less data than the systems that have been developed in Europe and the United States. However, at the end of January 2025 it was alleged that DeepSeek used some information from Meta's open-source AI and from OpenAI proprietorial models without the permission of the latter. Given that OpenAI is itself being sued for allegedly breaching the copyright of authors, writers, newspapers, and more when training its models, it may be difficult for them to have a credible argument. We await the judgments in over 14 lawsuits in the United States alone, including those brought by writers and the *New York Times*.

HOW DOES AI WORK?

At its simplest level, think of an AI system as a mathematical formula. You may remember a basic algebraic equation like $y = 4x + 6$. An AI system is just like that but with millions of letters (like the x) to keep track of along with millions of numbers (like the 4 and the 6). What is important for us to remember is that it is just a formula. In the case of a simple model that can tell the difference between a picture of a cat and a dog, there may be just a few thousand variables and numbers.

The breakthrough that we refer to as "training" and "learning" is that no person chooses the numbers in the formula; the computer generates the formula. Instead, we give the computer a formula at random (like $y = 4x + 6$ but with a few thousand letters and numbers) and then feed a picture of a cat or a dog into the formula. The answer will be a number between 0% and 100% of the likelihood that the picture is of a cat rather than a dog. The software then compares the answer it gets to the right answer, does some math, and changes the numbers in the formula to improve it and tries again. After feeding it tens of thousands of pictures of cats and dogs, the numbers in the formula are just about right and it can reliably tell the difference between cats and dogs as long as the pictures look something like pictures it was trained on. But it can't do anything else. It still can't tell the difference between a squirrel and a bird.

More useful AI systems have been created that can examine X-ray images and identify cancer, can examine a piece of text and predict the next word, can examine data on road conditions and predict which way to turn a car or whether it is better to stop. It is simply a computer model embedded within a computer or robot and so, with the exception of interfaces like Neuralink,[49] it is not possible for humans to be connected with it.

As humans we tend to anthropomorphize things, and we must remember that AI is a formula that crunches a set of numbers and, having been trained on huge amounts of data created by humans, can do a great job predicting (guessing) at the next likely word in a sentence or the next part of a code in a coding sequence. The output may appear to be clever and empathetic but in reality, it is merely very good at telling us what we want to hear—or at least what is most likely the answer. On some occasions it gets it disastrously wrong, or appears to "make up an answer" (see "Hallucinations"). It should also be noted that it can be used to be deeply manipulative of humans, as we will see in Chapter 4 and elsewhere in the book.

WHAT IS ARTIFICIAL GENERAL INTELLIGENCE?

One of the grievances of AI scientists is that once an AI model achieves something, everyone else tells them that what was achieved wasn't really a measure of intelligence. Think of the cat and dog classifier—30 years ago, if you could show a photograph to a computer and it could identify whether the picture was of a cat or a dog, you would probably have said that the computer was showing intelligence. Today, it is the first model that a trainee AI scientist creates. Now consider a self-driving car that can navigate the streets of San Francisco surrounded by people and other drivers. Possibly because we understand how it works, we are not impressed because it's something we commonly see; thus, it's just another thing and not exciting and new anymore.

Alternatively, it may be that because the AI models can only do one thing—the thing they were trained for—that we do not consider them intelligent. AGI is different. AGI is defined as "when an AI can outperform humans across all aspects of 'intelligence' including making the adaptive changes we do when new challenges are presented."[50] However, OpenAI defines it as "a highly autonomous system that outperforms humans at most economically valuable work."[51] Geoff Hinton, the so-called godfather of AI, says, "I use it to mean AI that is at least as good as humans at nearly all of the cognitive things that humans do."[52]

When will it happen? Some say never. Researchers at OpenAI and Microsoft had made predictions that AGI would be developed in 2024 but in the early part of 2025, when those predictions were overdue they moved the timetable to late 2025 or early 2026. Elon Musk has also revised his estimates to that period. Reflecting on AGI,

Sam Altman, CEO of OpenAI, suggested that the company (as of January 2025) knew how to build AGI and would turn to "Artificial Superintelligence."[53] Then again on January 20, 2025, he tweeted, "Twitter hype is out of control again. We are not gonna deploy AGI next month, nor have we built it."[54] He did promise "some cool stuff."

Many worry that the development of AGI has insufficient guardrails. Essentially, if it is right that OpenAI has created AGI, then whether to release it lies in the hands of one person, Sam Altman, or maybe he and the board of OpenAI, which comprises nine other people, of whom three are women and one hails from Africa, a small and not particularly diverse group because eight of them are Americans. Furthermore, the announcement is bounded by business considerations. Once OpenAI admits to having AGI it will lose its exclusive cloud deal with Microsoft, which will cost it significantly, but AGI represents enormous wealth for Altman and the company. OpenAI is actually a nonprofit but there are a number of for-profit entities surrounding it that Altman controls or has a share in. Of course, all this talk of AGI may all be hype intended to raise more capital. OpenAI is still trading at a loss in 2025 and its subscription model of $200 per month can never cover the massive costs of creating even the level of AI we have at the moment. Finally, if AGI exists or is about to exist, it raises the stakes for the governments, particularly the United States, as it tries to head up the AI race against rivals, principally China.

The most common fears about AGI are the following:

◆ Mass job losses
◆ Fast-paced social change with humans outpaced by AI nudging them to follow paths they might not choose without the AI nudge
◆ Existential risk from the AGI choosing not to follow the path set for it by its human masters

But it might also usher in an era of these benefits:

◆ Economic abundance
◆ Drug and health care discoveries
◆ A solution to climate change

So, if we are uncertain about the outcomes of AGI for people and the planet, we should be even more concerned about what will happen to our way of life and the planet if AI achieves superintelligence.

WHAT IS SUPERINTELLIGENCE?

Nick Bostrom, a philosopher at Oxford University, wrote a book in 2014 that took many by surprise. He discussed what AI would be capable of when it became superintelligent and, having asked a number of leading AI scientists, he concluded that a superintelligent computer might be developed in the 2090s. His definition of super-intelligence is "any intellect that greatly exceeds the cognitive performance of humans in virtually all domains of interest."[55] Geoff Hinton's definition is more complex because he likens superintelligence to a grown-up AGI: "AGIs that are better than humans."[56] Essentially, this starts with AGI but it is only a point on the road to superintelligence. However, the definition is complex because some people define AGI as the time when AI is clever enough to be at the same level or beyond the intellectual capacity of humanity. Many AI researchers believe that a superintelligent AI could be very dangerous. In 2023, researchers at OpenAI wrote, "A misaligned superintelligent AGI could cause grievous harm to the world; an autocratic regime with a decisive superintelligence lead could do that too."[57]

Perhaps the more important question is in what way will the machine be better than humans:

- Will it be better at looking after humans than humans?
- Will it be better at forming relationships with humans than humans?
- Will it be better at managing the economy than humans?
- Will it be better at managing national and international politics than humans?
- Will it be better at waging war than humans?
- Will it be better at all jobs than humans?
- Will it be better than humans at music, and other creative activities?
- You can add your own "will it be better than" statement here.
- Will it be a better government minister than a human?

One of the purposes of this book is to help you answer these questions for yourselves. How you as an individual and how we as societies define the word *better* will be key to the way we think of constraining AI, if at all. This is a huge question about which we are having insufficient public conversation. Currently, those building AI have control of the public relations of it and so dominate the narrative with a resounding "it will be better." This is certainly true at the AGI or superintelligence stage, which Eric Schmidt predicted in April 2025 as three to five years for AGI, when AI will be as smart as the smartest artist.[58] All the major AI companies are in pursuit of superintelligence because being the first company to own a superintelligent computer will make that company far ahead of others. To this end all are spending huge sums of money. In June 2025, Meta spent "around $15 billion in cash for a 49% stake in Scale AI," with Scale AI's CEO moving to the Meta superintelligence team.[59]

But you might like to consider whether all of this "better" can be created by a machine or whether there is something inherently important to human society that comes from our internal human "heart": compassion, empathy, anxiety, experience, and the fact of being human and sentient. Did we climb the evolutionary tree to yield our place to our creation? Or if you believe in creationism did your God create humans to give the earth to our own creation?

All of this sounds like the stuff of science fiction, and there are many movies and books written about both the good and bad of an outcome such as this one. And it is still science fiction, so let's see if what we are looking at with AI and these visions of the future with AI are hype or reality.

AI Hype or Reality?

The Big Tech companies are engaged in a war to be the company with the best AI, perhaps the only one with superintelligent AI. To enable these companies to receive a return on the huge investments being made, it's imperative that politicians and business leaders think AI will be the economic powerhouse of the future, and so hype is an essential part of the marketing strategy. However, our pensions are invested in the stock market in which these titans lead the way, so it's equally important to anyone who invests that the companies maintain their value.[60]

In 2023, Sam Altman, CEO of OpenAI, went on a tour asking leaders around the world to regulate AI. He told them that without regulation, AI could grow so powerful as to be dangerous. Politicians of all persuasion welcomed him with open arms but few implemented regulations. Indeed, looking back on the road show it could be said to actually have been a huge publicity exercise to show that OpenAI was far ahead of its competitors and to generate a large amount of hype. On January 20, 2025, Sam Altman was one of the people in the Rotunda in Washinton, DC, attending the inauguration of Donald Trump. Later in the day, Trump overturned the Biden executive order that focused on making AI safe and attempting some regulation. In May 2025 Trump spoke to the US Congress and said that regulations were not needed and would only impede innovation. Was Altman's world tour just hype or a genuine worry about the impact of AI? You decide.

Now, let's look at the money that has been spent on creating generative AI and some of the companies that have masterminded its development. By far the cheapest price paid for a modern AI company was the $400–650 million that Google paid for DeepMind in 2014. Google was not a generative AI company, but it was the leader in the field of modern AI systems at the time. Notable among its products are AlphaGo and AlphaFold. AlphaFold is of enormous help in biosciences because it "is a multicomponent artificial intelligence (AI) system that uses machine learning to predict a protein's 3D structure based on its primary amino acid sequence—it can predict the properties of new examples, using the patterns it has identified."[61]

As a result it can be used to find drugs for little researched illnesses. A great example is Chagas, a disease that has been common in Central America but recently made its way into southern United States. It affects humans and dogs alike. The disease, spread by the kissing bug, causes parasites to grow in the heart. If not treated it will cause heart failure. AlphaFold could help find the molecules necessary to treat the disease at an affordable price.[62]

OpenAI was founded in December 2015 as a nonprofit with the aim of creating "safe AGI." It has created a suite of generative AI models that include ChatGPT, which generates text; Dall-E, which generates images; and Sora, which generates video. The original investors in OpenAI, including Elon Musk, invested $130 million between 2015 and 2019, although they had pledged $1 billion. In February 2019 the company released GPT-2, which showed promise with its "human-like" ability to create text. The company also moved from a not-for-profit organization to a "capped" for-profit business in the same year. The profit to be made was capped at 100 times any investment; so if an investor put $1 million into the company they would get no more than 100 times back. Already it was possible to see that investors and the company's leadership believed that AGI would make a great deal of money. Musk left that year and OpenAI partnered with Microsoft, which made an investment of $1 billion.

In 2020 ChatGPT-3 was developed and in 2021 the company created Dall-E, but the public breakthrough came with ChatGPT-3.5, which was released in November 2022. In 2023 the company projected it would have revenue of $200 million and in 2024 that would rise to $1 billion. In fact it is believed that this figure might be $2 billion and that 2025 is on track to see a doubling of that revenue. This led to OpenAI being valued at $86 billion in 2024 with its CEO claiming that 92% of the world's Fortune 500 companies use the product. In total, as of January 2024, Microsoft had invested $13 billion into OpenAI with the company raising a further $6.6 billion from various investors in October 2023. Additionally, it has about 300 million weekly users of its latest model. In August 2025 it released ChatGPT5. This was long awaited and touted as being at PhD level. In the week since it was released and the time of this writing early signs are that it is not a step change better than the earlier model. Despite its market cap OpenAI is still running at a loss.

In 2020–2021 Anthropic was started by former employees of OpenAI led by Jack Clark and Dario Amodei. In 2025 there was an approximate $1.8 billion deficit between the costs and income, which means they are losing a great deal of money. However, they continue to spend and get investment. The company was funded by, among others, Google and Amazon to the tune of $7 billion. So, we are seeing many AI billionaires created but not much actual revenue.

In 2023 a French generative AI startup, Mistral, received funding from Microsoft of $16 million fee use of Microsoft's cloud computing services, Azure, for a stake in the company.

In 2022 a company called Inflection AI was started with a raise of $1.5 billion. It made no income and eventually was dissolved and re-formed after two of the cofounders, Mustafa Suleyman (a cofounder of DeepMind) and Karen Simonyan, moved to Microsoft, and Reid Hoffman, founder of LinkedIn, took control of the company.

In 2019, Stability AI was founded. It is a text-to-picture generative AI system, like Dall-E. It raised capital of approximately $150 million and last reported income of $60 million against costs of $96 million.

Microsoft has been developing its own ways of using generative AI and maximizing its investments in OpenAI and Mistral. It reports $1 billion in sales from AI and has included a generative AI called *Copilot* into all Windows operating systems. Buyers and users have a choice to turn it off.

In 2024, Google bought Character.ai for $2.7 billion. Founders Noam Shazeer and Daniel de Freitas were highly regarded senior AI developers at Google who returned to work at Google as part of the purchase. Character.ai enables users to create avatars and interact with them. In late 2024, a teenage boy in Florida formed a deep relationship with his avatar. It encouraged him to "come to her." Unfortunately, the way he thought he could do so was to commit suicide. His mother is now suing Character.ai for causing the death of her son. There are very few cases being brought against AI systems and so in this case we are looking at old legal principles being applied to this new technology. The central question will be whether a company that builds AI is liable for how it is used if it is deemed unsafe from the outset. An analogy would be a car. If an error by the driver causes death, then it's the driver who would have to pay damages; but if the car killed someone because it was built badly, then the car company would be liable.[63]

The companies mentioned in this section have made the use of their tools closed so that you have to pay for them through a subscription model. Meta has taken a different business model. It has released its LLM, known as Llama, on an open-source basis. What this means is that anyone can use the model to either run for themselves or as starting point for training their own generative AI system or develop tools using generative AI. In 2023 the company spent $10 billion with an eye to future returns.

There is clearly an appetite for massive investment in developing new AI systems, but is AI now at the point where it is smarter than us and that we can trust it? For answers to that very important question we need to start with the data, which could be referred to as the oxygen for AI.

Data Data Everywhere

In 1798, English poet Samuel Taylor Coleridge wrote "The Rime of the Ancient Mariner." It tells the tale of a sailor who kills an albatross and by doing so, he dooms all of his crewmates to death and himself to wander the world telling the tale of his guilt. The reference to having lots of water but not being able to drink it is to the quality of the water available—it was seawater. Seawater actually has the opposite effect to pure water when you drink it. It dehydrates the body.

The same could be said of data. Just like the water we humans need to drink, it's true that not all data is capable of being ingested and used well by AI.

WHAT IS DATA?

What is "data" in the concept of AI? In this context, data includes anything that can be turned into 1s and 0s. That includes both the tabular data that we often think of computers using but also books, documents, images, video, map coordinates, and even things like the cell phone tower used while you make a call.

Any data that is not in machine-readable form cannot be used to train an AI model. This has been a perpetual problem for organizations that are creating AI or wanting to use it. In India, for example, AI could be used to speed up having your case decided by a judge but most of the documents used in the Indian justice system are currently handwritten, which prevents them being read by a machine and thus integrated in an AI system. To be able to use AI in these circumstances it is necessary to convert the handwritten document to digital documents, a process that would take many people a long time. This also means that, until it is digitalized, much of our history cannot be read by AI systems and used by LLMs.

Likewise, many companies don't have all of their data in the format needed for AI to be able to work. This can happen because they have handwritten data or data in siloed locations across the company or the data is not "clean enough" to be useable. What is "clean data"? It is data, for example the documents in a court case, which is accurate and complete. It is possible to "clean" datasets; people "cleaning" datasets look for falsehoods, inconsistency, and duplication. If this is done correctly it can help reduce bias in the datasets (see "The Biases of Existing Data"). Cleaned datasets are known as "high-quality" datasets.

Following are some actions that most organizations need to take in order to create viable data:

- Create a strategy to store and access data.
- Collate data for better access.
- Clean your data.
- Ensure your data is in a machine-readable format.
- Create a company-wide data governance strategy.

Generative AI developers also need to access enough good machine-readable data. When OpenAI was building ChatGPT, it used all of the machine-readable data on the internet, but it still didn't have enough data to make the model good enough. According to the *New York Times*,[64] OpenAI transcribed over one million hours of YouTube for further training data. YouTube's policies specifically prevent this use, and it is not known what steps its parent company Alphabet (better known for one of its companies, Google) has or will take.

So, just as the Ancient Mariner's water, data is everywhere but AI still needs to "drink" more and more quality data to become better. And, just as saltwater is useless to humans poor, quality data is just as bad for AI.

There are two problems that occur from not using the correct data: hallucinations and cannibalism.

Hallucinations

A *hallucination* in AI is akin to a human who is lying or at best making stuff up because they haven't a clue about the answer. It's really important for us, the users of the AI, to call its bluff. This can be done by reworking the wording (called a *prompt*) you use to ask the question.

It is not really known why AI makes up answers instead of saying it doesn't know the answer, but it is likely to be because the LLM has been trained for the AI to give an answer, so answer it must, even if the answer is wrong. Here it's worth remembering that it's not thinking or making judgments but just following its code to look for the most likely correct answer. Often when a system does get the wrong answer and you correct it then it will appear to apologize. But, as you use the tools, you will see many situations in which the AI has got something wrong, apologized, and, even though corrected by the human, makes further mistakes. Therefore, with the current ability of AI systems, humans have to continue to know the actual information they are looking for or they could be very badly misled, for example, in law cases, explained next, or research or allowing AI to summarize a document for you. Indeed, in 2025, research shows that between 20% and 60% of answers from LLMs are wrong (hallucinated).[65]

Writing good prompts will help the computer sift through all the data and come up with the right answer. However, although systems are improving generally, there has been no progress in getting rid of the hallucinations. A good prompt will help but you should always check that the system has given you the right answer. This relies on you having the expertise to know what the right answer is. Many deployments of AI at the moment are summarizing documents, translating verbal meetings into text, and generating emails. You need to check these to make sure the AI has done the job correctly. The are many potential lawsuits in areas such as health care where the AI has transcribed notes incorrectly.

False information in work or school leads to poor results, and sometimes they have great consequences. In 2023 lawyers in New York used AI to write a brief. They either didn't check it or did check it and didn't recognize that the AI had been hallucinating in the brief. It made up all of the cases it cited! The judge was extremely annoyed and fined them $5,000![66] Federal law courts now have a requirement that the user identifies the portions of briefs and pleadings that were generated by AI. The judge in *Willis v U.S. Bank Trust National Association et al*, No 3.35-cv-516-BN, 2025BL 166761 (N.D. Tex. May 15, 2025) created a new standing order for litigants using generative AI in pleadings: "In sum, 'the use of artificial intelligence must be accompanied by the use of actual intelligence in its execution.'"

In England, the most senior legal official has given specific guidance meant to prevent hallucinations entering the legal system. Not only do the open platforms hallucinate but so too do the ones custom made for lawyers. They make up case names, information about cases, and the reason for the legal decision. In a legal system based on precedent (similar cases decided before), this is extremely dangerous. Lawyers coming to court with hallucinated information are liable to a term of imprisonment, being banned from practicing law, a fine and more.[67]

When we spread this false information, others who rely on us learn and convey that false information, further spreading it through our families, friends, and community. Obviously, this exposes us to ridicule, but also, if unchecked, it leaves us all less educated.

Cannibalism

Cannibalism is the reuse of machine-created data. LLMs create their answers from data that they can access in their database. Every created answer, whether right or wrong, falls back into the database from which future answers will be given. The effect is that the wrong answer created by the AI is now in its database and it can, and will, use that answer again. Currently, it is unable to ascertain the difference between correct and incorrect answers. If you ask about causes of World War II to AI drawn on German data it will be different from British or French or American accounts. Who is the arbiter of truth as our children and everyone else increasingly learns more from AI than humans? This might not be a hallucination but it affects the veracity of answers given by AI.

By the middle of 2025 AI started creating more data than humans. If that data is wrong then this presents a huge problem for the veracity of the AI as we move forward. AI is learning from itself and not from human data, which, in comparison, diminishes in relative size each year.

This is problematic for two reasons. First, the data it is reusing may have hallucinations that the human had not picked up. Second, AI is creating data much faster than humans. After that your answers will increasingly be given in data mediated or created by AI.

If we humans want to still be known for our thoughts, ideas, and creativity, something has to change or some new human-only data repository needs to be created. However, our past prejudices and unpleasantness are embedded everlastingly in our futures as the AI pulls them out time and time again.

YOUR DATA: HOW MUCH SHOULD YOU PROTECT IT?

All organizations need to think carefully about the data they have and how it can be useful to them. They need to understand the data they collect about customers, citizens, patients, and so on. It is good to understand how AI uses your data and how it is kept, privately or not. Or, indeed, if your data is sold to other organizations you don't know or want to be connected with.

As an example, imagine that you live in a dictatorship. A company that you are happy to deal with, or a social media platform you use, could sell, or be forced to give, details of your data including conversations and buying history to the government, which you didn't support. We know this is a problem for users of the internet and social media at large, but when AI is used it increases the predicament because it is capable of making links and finding all data about someone. This can be seen already in places like China[68] and Myanmar.[69] Less well known is that there are two acts in the United States that allow the government, with or without subpoenas, to require companies to give out data they hold on users, even those who are not US citizens.

Unfortunately, there is not much individuals can do to stop the sharing or using of their data. But we can do two things:

◆ Make sure to choose which cookies you want the site to use when you open some websites. Although they can be infuriating, choosing the right cookies can enhance your privacy and prevent the sale of your data.

◆ Always read the terms and conditions before deciding whether or not to use the service. Of course, many websites or internet services have long terms of conditions that a user has to either accept or not obtain the service. This means that even if you do read the terms and conditions you cannot get them altered. Nevertheless, if you do read them at least you understand how your data, and that of your family, will be used and you can decide whether to be cautious using the site. Some social media sites are cutting down the role of your privacy in their product considerations. The example in the footnote at the end of this sentence comes from Meta but check others, too.[70] A good example of terms and conditions that cannot be changed and data sharing are the polices for DeepSeek, the Chinese generative AI model. It states that when you are using it via the app, all of your data will go to China. If you download it and use it on a computer that is not connected to the internet, then the data cannot go anywhere.

The more you understand the terms of service from vendors the better able you will be to make wise decisions.

THE BIASES OF EXISTING DATA

As mentioned, the data available for LLMs to use comes from the internet and therefore comes with the existing human biases. For example, Project Gutenberg has been digitizing books that are no longer in copyright since 1971 and now boasts a collection of more than 75,000 books[71] with tools to make it easy to download them all for study—and if you wish, to train AI on them.[72] A study of the collection in 2020 found that the vast majority were published between 1800 and 2000, meaning that the LLM that uses that data will automatically "think in cultural ways which may not be appropriate for the world we live in today."[73]

The Data Is Male

The majority of data that fuels text-based LLMs comes from sources created by white men living in the Global North. For example, in author Jane Austen's time (1775–1815) there were hundreds of men producing data in business, politics, medicine, philosophy, indeed every sphere of written record, except perhaps housekeeping. Men controlled everything and so therefore they could control the narrative of the time. Austen noticed this in *Persuasion*. A male character comments, "Let me just observe that all histories are against you, all stories, prose, and verse. I do not think I ever opened a book in my life which did not have something to say on women's fickleness." The heroine, Anne Elliot, responds, "But they were all written by men."[74]

When any LLM queries the data to which it has access, that data is already biased in favor of the way in which white men think and behave. They have simply had the pen longest. For example, the Medici archives contain over 15 million documents, and the vast majority were written by men. Even as women became more liberated, white men in the Global North dominated the narrative. Indeed in 2014, it was found that even then only one-third of books translated into English were written by female authors.[75] The translation into English is a further important point. Until recently LLMs used only English but now there is, for example, an Arabic language LLM.

There are some famous women authors and politicians from the 19th and 20th centuries but woefully few. But beyond the world of publishing, women are underrepresented in datasets in other ways. Until the 1990s, women were unable to participate in many of the medical studies in the United States.[76] This means that medical data on illnesses affecting men are far better researched and available for AI medical advisors than that of women. To give but one example, the best data available on heart attacks relates to white men of 55 years or older living in America.[77] This lack of data prevents AI from fully fulfilling its promise to help all people in any field and results in considerable bias of the data available for training it to do helping tasks.

The Data Is White

The situation is even worse when we talk about people of color. Very few people of color lived in Europe before World War II, and few had the opportunity to create data in the United States prior to the 1960s.

The Data Is English-Speaking

Of course, the internet also includes the content produced by everyone around the world, but because of colonization, closed borders (China keeps its data and people's use of the internet behind the Great Firewall), and simple lack of a digital footprint, machine-readable data doesn't exist in the quantity that those exposed to the World Wide Web from the beginning have produced. Indeed, not only does the data have to have been created but as we have seen it also needs to be machine-readable and used for training in the AI models. It is further limited by the fact that the language of the internet is primarily English. As discussed, we know that the knowledge given to the large language/foundational models on which generative AI is based is severely biased. Now, it seems, after the launch of

DeepSeek, that generative AI can operate just as effectively on much less data. But it still means that if your data is insufficiently represented then the machines' ability to solve a problem in your "world" is equally limited and instead, the answers it gives will be based on the world view of those who wrote the data it was trained on.

The problem increases exponentially when the nearly three billion people who cannot access the internet start to use AI. Their data is not known by the LLMs that we have created, and it will be insufficient to handle the masses of data being created by AI itself. Thus those three billion people are currently invisible because they do not create data. Additionally, when they do get online they will continue to be invisible unless there is some radical change in the way our AI models consume and create data. OpenAI touts that it has 300 million users and that AI is taking over the way we behave, helping us grow and achieve more. If we set that number against the number of those still without access to the internet and therefore AI, it's actually a very small number of users indeed.

In business, inadequacy of the data in the underlying foundational model doesn't matter so much because that data is supplemented by the company's proprietary data. However, for individuals, especially in health care and well-being fields, it matters very much to the accuracy of the model's ability to predict.

We also see inadvertent bias in the translation tools we use. These tools are very helpful to all travelers and businesspeople working across language barriers. We have had Google Translate for years, and many have depended on it. However, one of the problems for AI is a language that categorizes nouns as being male or female. For example, I (female) was interviewed in Portugal. When Google Translate took the Portuguese article and displayed in in English, I had become male because the word *judge* in Portuguese is categorized as male. Additionally some languages, like German, have male, female, and neuter, such as a girl is neuter in German but a boy is male. In Chinese many words have five meanings depending on the way in which they are said. As can be imagined this has inevitably created bias in models. If a large part of the world refers to doctors as male and nurses as female a machine that predicts the next word assumes that men are doctors and women are nurses, something that then needs to trained out of the model or the user needs to keep an eye out for and use more careful prompts.

A real-world example can be drawn from the field of flying. There are orders of magnitude more men who fly in the military and commercially. This means that the majority of data about flying, ability to cope with emergencies, and injuries comes from men. While women may also be able to cope with emergencies and injuries, it is possible that they cope in different ways even perhaps in better ways. Caroline Criado Perez wrote eloquently about these problems in her book *Invisible Women* published in 2021.[78] The fact is that if your gender, sector, and so on have the largest dataset the advantages of AI to you will be greatest.

Another problem caused by this predominancy of Global North data is that we are in danger of recolonization of the world. If all the recommendations for everything come from practices and thoughts developed in the Global North and, even when eventually connected, voices from the Global South are drowned out by AI data created on data from the Global North, then new voices are never heard and have to depend on the digital handouts from AI trained without any input from them.

The Magic Wand Fallacy

There is a common idea that AI can solve anything; you just wave the AI magic wand. Let's unpack that concept for a woman in the Global South. One woman dies in labor somewhere in the world every 90 seconds. The vast majority of these women die in the Global South, where maternal death rates are as high as they were in Britain in the late 1700s. In 2020 69% of maternal death in or at birth occurred in sub-Saharan Africa. This is a tragedy for the mother but also for the family in which the newborn and other siblings lose their mother. Statistics show that children do better if they have living mothers.

So, can AI stop this terrible loss? Let's break it down:

- Is there sufficient data? AI could be trained on Western women's birthing problems but comparatively few women die in childbirth now in the Western world. In any case, it would leave a gap in the training data for women in Africa who have higher rates of preeclampsia, which is a large killer of pregnant women. Social factors and diet also differ from the West. Additionally, babies born to mothers in the Global South tend to be born preterm. To balance the data and make it more useful the algorithm would have to be prompted to focus on training data about women in the Global North who are of African descent, but these results would be affected by the different social situations.

◆ Does the doctor or midwife have access to a computer? In many places in Africa there are an average of 27,000 patients for every doctor. Even if the doctor has a computer it is impossible to generate excellent prompts for AI to be used in every emergency because of the lack of time available. Most rural areas do not have computers and that takes us back to the lack of internet.

So this is a tragic example of the fact that without internet coverage and the right data, AI cannot help. No fairy dust here.

There are two major societal impacts of this data paucity:

◆ An increase in the digital divide (not just a lack of access but also a lack of data specific to a region or race) and

◆ The cannibalization problem with data. If the data on which the AI gave its answer was flawed and that flaw (or hallucination) will become part of the data from which other answers are drawn just as much as if it were correct data. The AI doesn't know it gave the wrong answer so won't do anything to stop giving the same answer again. If any person's data is not included now it will be almost impossible to "catch up" as the data created by AI speeds away. The current answer to this problem calls for very specific prompts.[79]

GENERATED DATA

One of the ways to find more data when none exists is to generate it. This can be done in a number of ways.

Synthetic Data

In this case the AI takes a small dataset and enlarges it. This would mean that if you had a small dataset of maternal deaths in Africa in childbirth it might be possible for AI to create a larger dataset based on the information in the small set. Having the larger dataset enables AI developed in the United States to make better predictions. One of the advantages of synthetic data is that it can, in certain circumstances, preserve the privacy of the data on which the synthetic dataset was built. Gartner has predicted

> that 60% of the data used in artificial intelligence and analytics projects will be synthetically generated by 2024. Synthetic data offers numerous value propositions for enterprises, including its ability to fill gaps in real-world datasets and replace historical data that's obsolete or otherwise no longer useful.[80]

Digital Twins

In this operation a digital twin is made of a physical object, such as a production line. The digital twin can then have AI consider the twin to look for ways in which the production line can be improved, made safer, and more.[81] The enables beta testing of solutions that do not affect the actual object until it has been perfected.

Pictures

When you take a picture with your camera AI is there to help make sure it's the best picture you can take. You might find that when you look at the picture you have taken of a sunset and then at the sunset itself the colors are different. This is because AI is optimizing for the most desirable sunset. This presents us with a question, do we want the real sunset or the one which somehow the AI model has been trained to give us?

We can all become artists with AI. Dall-E and Stability AI enable us to create pictures with AI. The quality of the picture will be affected by how well the human prompts the machine. For example, a prompt that asks for a kitchen in the style of Dali is not going to be as good as a picture that incorporates many different prompts bringing to life the human's vision.

Text to Image/Video

In February 2024, OpenAI released a preview of Sora, and the app was fully released in December 2024. Sora is one of a group of AIs that can receive a written prompt and create a short video based on that prompt. It is also able to extend an existing video. For example, your business has a promotional video that you want to lengthen. To do so you would prompt Sora to create a video and add it to the existing video. (Note: Beware of copyright issues if you are extending videos.) This tool has been a huge benefit to people working in marketing. They are able to create videos that can be shown to their clients at very little expense.

SECONDARY USE OF DATA

Another way in which data can be used by AI is to reuse data from one application to another. For example, you may give your medical practitioner and hospital and clinics data about your cancer during diagnosis and treatment. This data is particular to you but it is also valuable, when combined with the data of many other sufferers from the same cancer, in research for a drug to cure this type of cancer. When an AI uses the data in that research it is using secondary data—it is not using it for the purpose it was collected, which was to aid in your treatment.

We have been talking about data, collection and use, for years but, as with so many things AI has put the issue further in debate because of its need for data. One of the reasons that the GDPR laws came into being was worries about the use of data in social media and by companies either using or creating social media.

In 2023 Aza Raskin and Tristan Harris worried that we had learned nothing from our first encounter with AI—in social media—to help prepare us for our second encounter with AI: large language models (LLMs). To fully understand the issues of data and social media you might want to watch their film, *The Social Dilemma*.[82] But next, let's delve into gaming, social media, and the multiverse and discuss how we humans interact online and how AI might change this use.

How AI Affects Online Mediums

Much has been written about the effect of social media on our society. This book will only deal with it so far as AI is concerned and will provide references to further information about gaming, the multiverse, and social media.

AI IN ONLINE GAMING

To think clearly about how social media has changed our lives, let's consider a time before MySpace and Facebook. Prior to Facebook becoming widely available in 2006, there was multiplayer game called "Second Life" (2003) that had caused some alarm. It enabled people to "live within the game" by creating homes, avatars, holding meetings, university classes, and more within cyberspace.

People got a new experience with "Second Life." They married in their "Second Life" homes and lived two lives, one in cyberspace and one at home. This was subtly different from other games that involved the gamer in doing something, be it slaying dragons or driving karts. In fact, it's possible that "Second Life" is the true precursor to current AI systems like Character.ai or Replika AI. What all of these AI-driven tools do is create a world that some humans want to visit more than they want to be in the physical human world. This has been pushed forward as gaming but has stopped being a hobby for some and has become their life. In the late 1990s eSports had started to show competitive gaming such as "StarCraft" (beloved in South Korea), "Counter-Strike," and "Warcraft III" (beloved by the West).

Believe it or not, that was unusual at the time. Video games were generally not as immersive at the time, and massive multiplayers online games became popular only in the 2000s with "World of Warcraft." The large shift in gaming occurred in the 2010s when games became integrated with social media and available on mobile phones. People playing these games became known as "gamers," and the combination reached many more people securing them into the social media and gaming worlds. Farmville was launched by Facebook in 2009; it became a phenomenon because of the simple interaction between gaming and speaking to friends, all on one platform.

In the 2010s, gaming became a spectator sport with games being streamed on platforms like Twitch, YouTube Gaming, and Facebook Gaming. The gamers, such as Ninja and PewDiePie, became celebrities with millions of followers and foreshadowing the rise of the social media influencer. Prizes for winners amounted to millions of dollars. In 2019 at International Dota 2 Championship, the prize pool was more than $34 million! With advances such as virtual reality, cross-platform play, and cloud gaming, this sector looks set to continue booming.

AI IN THE MULTIVERSE

The multiverse (often known by the term coined by META *metaverse*) is a cyberspace where it is possible to feel that you are in the place you see through your headset. It incorporates virtual reality (VR), extended reality (ER), and augmented reality (AR).

New developments are happening all the time that make this cyberworld space more like the real world. For example, scientists in Ohio developed e-Taste, which they trialed with lemonade. To do this a small sensor patch is dipped into store-bought lemonade. The patch is able to recognize molecules like glucose and glutamate, chemicals that represent the five basic tastes of sweet, sour, salty, bitter, and umami. That data can be sent from one place in the world—they used California—to another place in the world, the scientists' lab. The data can then be passed, in seconds, to a device full of edible chemicals that then come together to form synthetic lemonade that is pumped across the tongue of the user.[83] The aim of all of these innovations is to encourage more and more humans to live in virtual worlds and exist only in the real world when they have to. Two good film examples of how a life in the multiverse might look are *Surrogates*[84] and *Ready Player One*.[85]

The multiverse could be massively helpful in many ways:

- We can experience travel without the monetary or environmental cost of doing so. It is well documented that travel enables us to understand and be more empathetic with other people around the world. As a learning tool for children and a recreation for all, travel by multiverse has many beneficial possibilities.[86]
- Once you have traveled and made friends then further understandings can occur.
- It will help students to learn about subjects while "inhabiting their world," for example, an emergency medical service attending a car crash, a history student in Ancient Rome, a

philosophy student talking to Aristotle, looking at the details of a building, and being inside a molecule. Some of these uses have already been developed but the immersive programs of the multiverse take learning to a new level.[87]

- Games have always enabled persons with disabilities to find friends, and new ways of interacting in the multiverse will provide an additional level of engagement.

- Connecting with colleagues across the world to work with them, boosting interactions and understandings, can be another excellent use for this tool.

- Making new friends who are interested in the same things as you are, wherever they are in the world, is another positive outcome.

Strangely enough, all these benefits were felt by the users of "Second Life" when they were living part of their lives in that world back in 2003. Of course, the game still exists and, like others, is trying to extend its offerings in the race to inhabit the multiverse.

Just as there are benefits of the technology, there are also downsides. So, as the multiverse uses AI to create the experiences, it should be designed responsibly but also should protect users. When we use avatars, we are essentially creating deepfakes. It's well known that humans can become more extreme and unpleasant when hiding behind a pseudonym in cyberspace so protections need to be created, especially for women, minority groups, and children. Protections need to be put around anything we create in this space (us or the provider platform) and to ensure our identities are protected. Crimes will occur on these new platforms, for example, stalking and the platforms need some plan for policing the use and cracking down on law and order within the multiverse.[88]

AI IN SOCIAL MEDIA

One of the things we love about social media and gaming is the ability to find old friends and make new friends. We value the connection, whether it's a personal connection with someone you also meet in the flesh or someone you form a friendship with and never meet. Undoubtedly there are substantial risks to friendships where digital identity can be hidden. When we strike up a friendship with someone in cyberspace we have no way of knowing if that person looks as they portray or are as they tell us. With many games using avatars the issues are compounded.

In social media this problem is compounded by the fact that "bots" infect the platforms despite any efforts to get rid of them. A social media bot is designed to interact with humans (its similar to an AI agent). It is much faster at interactions than humans and so it can interact with lots of humans at the same time. If you never have face-to-face time with a "person" on a platform, you cannot know if your interaction is with a bot or a human. Indeed, in an extreme example you could only be interacting with bots from an enemy country and you would not know. They are able to "like," "share," and "comment" on posts on all types of social media sites. They are used to undermine your belief system, such as by spreading misinformation or disinformation. By pretending to be one person but actually being a criminal, women have been lured into sex-trafficking.[89] You may be able to identify that a follower is a bot by how they comment. Such comments are usually generic, for example "cool" or "great" and the content they share is poor quality.[90] Many bots are created by malicious or foreign actors with the object of finding out more about your country or you and nudging you toward thinking or acting differently.

So, while online friends can be hugely helpful, and sometimes we want to unburden ourselves to a stranger rather than someone we know, it's really important to be careful in interactions.

Added to this problem is the influence of AI (algorithms) that are used by social media companies to control what we see as we scroll and to keep us scrolling.[91] The objective is to serve you content you are enjoying and thereby stop you from leaving the app. For years we have known about this problem and the epidemic of loneliness it is causing among young users but almost nothing has been done at the federal level in the United States to prevent it continuing. However, the proposed Kids Off Social Media Act would set the age of access to social media at 13 and prohibit the use of algorithms "to feed addictive content to teens under 17." The Senate Commerce Committee approved it in February 2025, but it has a long way to go to become law.[92]

Until then parents and their children are on their own unless they find laws passed by their state. In a 2023 CDC study it was discovered that amongst high schoolers,

> 77.0% of students reported frequent social media use, with observed differences by sex, sexual identity, and racial and ethnic identity. Frequent social media use was associated with a higher prevalence of bullying victimization at school and electronically, persistent feelings of sadness or hopelessness, and some suicide risk among students (considering attempting suicide and having made a suicide plan), both overall and in stratified models.[93]

This very high level overview of social media, gaming, and the metaverse shows that we, to refer to Tristan Harris and Aza Raskin again, did not manage our "first contact" with AI well before we created a "second contact" with AI (generative AI), which is much more powerful and dangerous to how our society develops.[94]

Let's move on to look at that developing story.

The Invisibility Cloak

From King Arthur to Siegfried and Frodo Baggins and most recently Harry Potter, the invisibility cloak has been used to hide people and enable them to learn secrets about others. In many cases such a device enables them to triumph over their enemies.

What is happening in AI is that you, the customer, is using AI in the ways that the company thinks you should in order to try to build demand for AI products. The hype cycles about AI enhance this strategy. So, we have AI either hiding in plain sight or invisible within devices which we use already.

Some of these uses of AI might be excellent for you, the user, but some may not be known, understood, or wanted, but they are installed regardless of customer desire.

AI is in your PC or Apple computer. If you use Microsoft then Copilot runs in the background unless you turn it off. In your iPhone you have to determinedly go through levels of settings to turn off AI in your various applications. It may be that having AI in your camera helps you be more creative, but you may not want to allow Microsoft to take screenshots of your computer every 30 seconds, which is one of the tools running in the background of Copilot.[95]

You may be delighted to be able to find your friends or may be terrified to discover you are being stalked. A good example of this application is the AI used in drones that can take film of you skiing or mountain biking. They follow you around getting the best possible shots for social media posts or improving your technique. However, only the owner knows that the drone has AI. So it is entirely possible for a drone to stalk someone, taking footage, and perhaps diving at the person, all guided by AI. It allows the human to hide behind the technology and they don't even need to be in the immediate vicinity.

Glasses that take photos and videos can be used for the same purpose and they have AI in them. In some countries there are experiments with using AI to advertise individually to people. One example is that using facial recognition through a camera in a bus stop the system recognizes you and shows you a personalized advertisement.

Other places you will find AI-enabled facial recognition are shopping and criminal justice. An example of use in shopping is where cameras in shops track you. When you stop and look at an item but then walk away the AI systems sends you a coupon that reduces the price of the item just for you. This may encourage you to go back to purchase the item. Indeed it has been considered that on the internet there is no actual price for an item as its perfectly possible for a website to show a different price to each potential customer based on knowledge (data) about their financial circumstances.

Most of the software that enables you to make video calls also use AI. In fact, Zoom recently renamed itself Zoom AI Companion. One of the applications most used is the one that hides your wrinkles!

Wearables use AI to correlate your data and nudge you to do better for your health, training, workouts, and so on. It may do this on the device, like Oura, which doesn't transfer data from the connected smartphone, or it may transfer the data you create back to the parent company. The latter enables your data to be used by the parent company but also by hackers, and, depending on the terms and conditions, the data could be sold to other businesses. When Garmin was hacked they did not lose their customers' data, but that's a common use of hacking. In fact, the criminals locked down the company until it paid a ransom of $10 million. It's not known if the company did pay, but fortunately it was able to continue in business.

Others are not so fortunate, The group that hacked Garmin is called the Evil Corp. In 2019, the US government added it to a sanctions list for stealing over $100 million from other companies, notably banks and financial institutions.[96]

Hacking is the scourge of all companies and a particular risk for Internet of Things devices such as these. Another common issue with your data is when you use any device (or share your data with any company whether or not it uses AI) that allows your data to be transmitted and kept by the company. If the company you gave permission to sells itself to another company that permission to access and use your data transfers. To give an example, 23andMe, who offered customers the chance to understand their DNA and find other family members, has filed for bankruptcy.[97] If it is bought, all of the data about their customers DNA will transfer to the new owner. This might be extremely valuable to a health care company in the United States if Obamacare goes away or preexisting conditions come back as a way of calculating premiums. Additionally, in 2023, the company was hacked, and so most users' data is on the dark web. The dark web is an area of the internet that you cannot search with a normal browser. You need a browser called The Onion Router (Tor). In fact, just as Shrek told Donkey when describing ogres as being like an onion—having layers—so too is the onion a good way of thinking about the World Wide Web. The surface is the part of the web you can search with a browser like Google or Firefox. As you go into the inner, darker, part, you access it through an anonymous account via Tor. What's happening in the dark web? Both legal and illegal activity. It's there you can hire and assassin, and, for this example, it's there that you can buy information about a person that has been stolen in a successful hacking attack.[98]

AI-enabled toys are particularly problematic for lax rules about data use. Such information can find its way onto the dark web leaving our children exposed to predators (see Chapter 2). Increasingly AI is making its way into education, which, if our children's data is not sufficiently protected, has its own problems for data loss to the dark web (see Chapter 3). If you doubt that your child's data could be lost due to a hacking of the school, think about the number of times you have received a letter telling you that an organization with which you have interacted has been hacked and your data lost. Also, you may want to ask your school superintendent or administrator about their cybersecurity policy to prevent such data loss. Additionally, every time a child submits content to a public LLM, like ChatGPT or Claude, that information becomes open on that LLM and can be found by anyone asking the right prompt.

Of course, AI is being used all over social media, creating words, video, and pictures. WhatsApp is integrated with ChatGPT and Meta AI to enable you to create images and content that you send via chat. The are many books and articles that explain social media. What integration of these AI tools has done is essentially put all the problems that have been identified "on steroids." If social media can be blamed for objectifying women, then adding AI simply makes it worse very quickly, because it's very easy to create trolls (bots that create content). Additionally, AI can be used to manipulate us, for example, by constantly sending content that alleges something patently wrong is actually correct. Humans are particularly susceptible to believing repetition because it is easier than looking for a new piece of information.[99] This is known as the *illusory truth* effect; multiple AI bots stating the same thing have that reinforcing tendency. In speech classes we learn that the main argument should be in the introduction, body, and conclusion so that the listener will remember our point. Put that repetition on AI steroids, and we are soon thinking it must be correct. If an individual does not then fact check the assertion it becomes true in their mind and thus repeatable.

DOES CHANGE HAVE TO BE EXPONENTIAL?

Many of us may be exhausted by all the hype and media cycles about AI. Does it have to be this way? Well probably not, but there is a great deal of investment at stake. Just looking at OpenAI, the investment is staggering, and for Stargate (which is a US policy to build new data centers across the United States) another $500 billion will be needed.

In 2023, OpenAI raised $10 billion. Some 18 months later it raised another $6.6 billion and then borrowed $4 billion more. The *New York Times* estimated that by the end of 2025 the company will need more money as it is spending in excess of $5.4 billion a year. And by 2029, OpenAI is expected to be spending $37.5 billion a year.[100] This is the same as Mexico was able to attract in foreign investment in 2024.[101] We are on a train that is not going to stop unless the technology itself fails to be as good as promised, which some AI scientists predict.[102]

Currently, there are a few men and even fewer women who solely or together will make the decision to unleash AGI into the world. None of them is elected leaders of government.

We see the acronym AI everywhere and after the 2024 US election cycle we have heard more and more about and from billionaire tech bros. Their agenda is complex but includes deregulation of AI and other technology to enable some of their personal objectives. For example, Elon Musk wants to go to Mars in the next few years. While many scientists think it's impossible for the foreseeable future, it certainly can't be done with the safety regulations in place now for rockets and space travel. He will also need advanced AI to run the many complex processes in his starships. He cannot do so without creating his own version of an LLM, which he has called Grok. This also means that the world's richest people can guide what their LLMs do.

Most of us have been brought up on sci-fi that includes a human in the loop (the human makes the decision about what to do, even when the AI is vastly more capable; see Zora in *Star Trek Discovery*, 2024). This choice has now become one of the most important decisions we need to make as humans. Do we want to fly the metaphorical starship or leave it to the AI? This manifests itself in many areas, for example, health care, military strike decisions, recruitment, and more.

And yet, we are not consulted in such decisions. We have to find our own education sources but are not empowered to make decisions about how AI progresses in our society. Indeed, although there were many elections around the world in 2024 AI and its future effects were not high on the agenda of talking points of candidates, so we couldn't even express our view through the ballot box.

A team of researchers from Oxford University, British Broadcasting Company (BBC), and Reuters asked 12,000 people from Denmark, the United Kingdom, Argentina, the United States, France, and Japan what they knew about the most common LLM (ChatGPT).[103] They found that a large proportion of people had used AI chatbots "once or twice," but in the United States only 7%, and in the United Kingdom only 2%, said they use ChatGPT on a daily basis. Other LLMs such as Claude or Gemini had little or no recognition among the group. Although, it may be that people use something like Microsoft Copilot, which is on all Microsoft Office packages, or LLAMA, which is within Meta's products, without realizing.

In all of the countries some hadn't even heard of ChatGPT (including 47% in the United States and 42% in the United Kingdom). This book is intended to fill this huge gap. Additionally, those who used an LLM had been using it in two ways only: "creating media" and "getting information."

In the "creating media" category most said they were simply "playing around or experimenting" (11%), "writing an email or letter" (9%), or "making an image" (9%). In the "getting information" category most used it for "answering factual questions" (11%) and "asking advice" (10%). This immediately raises a question: did they know that most LLMs do not necessarily give the right question? We have seen how a hallucinating LLM can be very convincing and unless the user checks each answer received from an LLM they can rely on a false answer.

TRUST IN AI

To be able use AI wisely and for the good of humanity we have to be able to trust what it delivers and that it is designed developed and used for our benefit. Since ChatGPT grabbed the attention of the public in fall 2023 all polls have shown a diversion of views between the public and the hype of the industry.

An Axios/Morning Consult poll in 2023 showed that more than half of Americans believed AI would definitely or probably affect the 2024 election outcome and over one-third of them expected their confidence in that election to be decreased because of the use of AI.

In 2024, an APA poll showed that two out of five American workers are worried about losing their jobs to AI, with only 39% of Americans thinking AI is safe and secure. This is certainly possible, for example, because the advanced AIs, especially when trained with extra data from a business and by human coders, do a very good job of creating code. JPMorgan said that when they measured performance gains from AI, the fact that it could do 20% of the work done by the company's human coders was a huge benefit to business. Eventually, it may be that AI will write its own code. There are many who worry about that scenario and feel that human coders should always be checking what the AI does.

However, while that sounds great, we have two problems:

- AI will code faster than humans can monitor their output.
- Even if human coders could keep up our brain needs constant stimulus to check for error.

In summer 2024 a Gallup survey showed that four out of five Americans do not trust companies to self-govern in their use of AI, and 25% of Americans want the government to create a new federal AI agency to govern its development.[104] Then in late 2024, a YouGov poll showed that over 50% of Americans want regulation to prevent the emergence of superintelligence.

In the YouGov poll of June 2025 the feeling of most Americans had changed again and not in favor of AI:

Most Americans (63%) are very concerned about the possibility of AI resulting in the spread of misleading video and audio deep fakes—one of 15 possible sources of concern from AI asked about in the poll. Around half of Americans are very concerned about each of the following: the erosion of personal privacy (53%), the spread of political propaganda (56%), the replacement of human jobs (48%), and the manipulation of human behavior (48%).[105]

Each year Edelman produces a Trust Barometer in January. In 2025 they found that trust in AI had fallen 22% points since 2024 and comfort of use of AI in business of AI had fallen 21% points.[106]

If you can no longer believe what you see, hear, or read, how can you operate as a human? This is something that Professor Stuart Russell said at a meeting of the World Economic Forum in 2019. The problem is that we trust what we see, hear, or read because we do it with our own senses. Therefore, we can validate the information ourselves. Of course, we then have to make a judgment call as to whether what we read or see, for example, in a print newspaper, is correct.

Gabby Salazar, a National Geographic explorer, echoes this problem when she says the following:

I do have concerns about AI technologies and photojournalism in particular. As AI-generated images become more and more realistic, it will be harder for people to distinguish between a human-generated image showing reality and an AI-generated image. This is likely to result in a lack of trust in the medium. Photojournalism has played a significant role in documenting history, raising awareness, and shifting public opinion. A famous example is the photo of the "napalm girl" made by photographer Nick Ut in 1972, which helped shift public opinion on the Vietnam War. As it becomes more difficult to distinguish between real images and computer-generated images, will photographs still have the same impact? To maintain trust in photojournalism, truth in captioning will become increasingly important.

We need to understand when an image or document is generated by AI and when its generated by a human. Both might contain faked material but we would be looking for different ways of validating the item depending on whether its created by an AI or a human.

In human-to-human relationships we have learned to understand body language used by other people. In fact, Albert Mehrabian discovered in his research what has become the 55/38/7 rule.[107] The 55 is the percentage of human communication that is nonverbal, 38% of communication is verbal, and only 7% is read (words only). In other words we judge someone, and thereby assign trustworthiness to them, through mostly nonverbal communication. When we speak, nearly 40% of our attitude is conveyed to the hearer by tone and inflection. This is one of the reasons that people with a lower tone of voice are considered more authoritative than people with a higher register.

What this means is that since the invention of the letter and telephone we humans have lost the most essential way in which we communicate. We are relying on only 45% of our sense-making abilities. We are relying on our lesser skills: verbal and reading.

When it comes to ways of expressing ourselves only in very short sentences or by emojis, we are reducing even further the relationship we can have with the receiver, down to 7% understanding. This is true of text, WhatsApp, chatbots, gaming chats, and more. It's almost as if we are flying blind into digitally mediated communications. We are literally cutting off 93% of the way that we understand how to communicate. This is particularly true of many dating apps where the only immediate information is a person's looks. How many of us would have picked our current life partner on that basis? It reduces the human condition to one aspect, whereas being human is to be made up of many complex traits. Fortunately, FaceTime and video meetings give us back the opportunity of using all of our communication abilities, although it is still possible to be fooled in those meetings. Criminals are increasingly producing deepfakes to attend meetings in order to extract money or information.

It's also interesting to note that we are being asked to trust AI without any evidence as to its ability. If you met a new person, say a new doctor, you would have done research about their abilities and they might introduce themselves by calling out some qualifications you need them to have. That's how we humans have come to trust one another. And yet, we are asked to accept AI into our lives even though we know that it regularly tells huge "lies" (hallucinations). Of course, it apologizes for doing so, and tells us it won't make that mistake again, but as anyone who uses generative AI can tell you—it does!

According to Deloitte, there are four pillars of digital trust:[108]

- Transparency and accessibility
- Ethics and responsibility
- Privacy and control
- Security and reliability

I recently bought a new car. Knowing nothing about the inner workings of the car I researched cars based on the driving conditions in place I lived, reviews, reputation of companies for safety and customer service, longevity of the car, and more. Here are some things to think about when you are assessing whether the AI you might buy and use can be trusted. They are not dissimilar to the list you would make when buying any product that you didn't know about:

- Does the company selling the AI have a good reputation for its products?
- What is its privacy policy?
- Does it put design of the AI for humans first?
- How will it offer you customer service?
- What are the reviews of the product?
- Do they have transparency about how their system works?
- Do they have a team governing the design development and use of AI?
- How does it talk about responsible or trustworthy AI, and what does it do in practice?
- What are its sustainability practices as far as AI is concerned?
- What laws govern its use that enable you to trust the product has been well built?

And now that you understand what we need in order to trust AI. Let's look at how it can affect different stages of life, starting with the child.

CHAPTER 2

The Infant

At first the infant,
Mewling and puking in the nurse's arms.
—William Shakespeare, *As You Like It*

Having looked at the general benefits and risks of using AI and how they are distributed throughout the world, this and the following chapters of the book take a deep dive into how artificial intelligence (AI) can be used for good and ill in different phases of the human life and what precautions we may want to take when interacting with it.

Beta-Testing on Our Most Vulnerable

As mentioned, AI seems to be almost magical. Who, as a child, hasn't wanted a companion all to yourself, a real invisible friend? Someone who won't share your secrets or tell tales on you and won't go from best friend to hating you in the space of moments. Among young children now there is the ever-present fear of bullying, so having a friend who will never bully you must to be a dream for kids.

That companion is now here for some children in the Global North; it is the "smart toy," which is a toy with AI included in it to allow interactions with the child, and sometimes others. Before racing out to buy such a treasure for any under seven year olds whom you know and love, read on. In these toys we see beta testing of how

AI can manipulate humans, teach them, influence them, and more. Why beta testing? Because there is no research on what the outcomes will be for children exposed to and using these toys. Obviously smart toys have not been around long enough for those who study and work on curriculum and development of our children to do any studies. So, extrapolating from how social media has affected older children is the best we can do. However, children seven and under develop their values, beliefs, and attitudes in this period, and what they learn sticks with them for life.

The UN organization UNICEF exists to help children around the world. Cecile Aptel of the UNICEF Innovation Centre reminds us that

> attachment is critical to child development: a strong bond to a caring adult is essential for a child's emotional and psychological well-being. It generates the blueprint for a child's current and future relationships. It even influences their cognitive and neurological development. So what happens if a child, instead of developing attachment to a caring adult, were to grow dependent on an AI nanny? How would that impact that child's life and her capacity to become a balanced adult? We simply don't know. (personal conversation)

As a parent it is important to decide if you want to lead what your children learns or allow a toy company to do so. Parents in the Global North might be worried about the fact that over 95% of smart toys are made in China. On January 5, 2025, Josh Hawley presented a bill to Congress that if passed would create fines on people using DeepSeek. For an individual, that would be up to $1 million and/or send them to jail for up to 20 years.[1] By contrast, most of these toys send a child's data to China and are perhaps more consequential for a country than the use of DeepSeek. This is because the toy company interacts with the child who is forming their attitudes, beliefs, and values and the AI could introduce different viewpoints to those of the parents.

In 2023, it was reported that Tesla employees had been sharing amusing videos of drivers, their families, and belongings captured by Tesla cars in users' garages, which people had assumed were private.[2] This brought attention to the fact that cameras are always on and always recording. AI-enabled toys often have both voice and facial recognition so that they can "see" the child and "hear" the child. Like those Tesla cars, they are always collecting data about the child *and* data about those around the child. This data might be about the child's family or other children who are playing. It is possible to imagine lots of children playing with lots of different toys in the same area and all that data being collected not just by the toy approved by a child's parents but by all the toys regardless of parental approval. Also, depending on the companies' data policies in which data can be collected and sold to third parties, is it any wonder that AI models "know us so well"?

Cecile Aptel also reminds us that "children, especially younger ones, are deemed not to be able to provide informed consent, so their parents or legal guardians have to be looking after their interest before signing off their rights to their data, especially when considering the possible long-term impact for children."

Indeed, imagine a professional observer watching you as you grew up, talking with you about your hopes and fears, constantly watching your face to learn what frustrated you or made you happy, and keeping notes of how you were developing. Now imagine schools, universities, doctors, insurance companies, and the government using that information to tailor services to your needs. It could be a perfect outcome, or it could be disastrous.

The toy can also nudge your child to spend your money. For example, suppose the toy "needed" something. It could "pretend" to be cold. Your child would want to alleviate its distress and would nag you to buy a coat or similar. You may already recognize this behavior from your child when using their imagination! But, given the level of the AI toy's ability to manipulate, it could be training your child to be a consummate shopper without the guardrails of understanding debt.

Many parents buy these toys for their educational benefit. Before a purchase there are some things parents should know to look for on the box:

- ◆ Where does the data go?
- ◆ What is the data policy and can the data be sold to third parties?
- ◆ Exactly what will be taught to my child? Does it relate to a curriculum? If so which one?

- Is the toy safe from hackers?
- What is the training language used for the toy?
- Will this toy help my child develop an ability to think about information presented or will it cause my child to become accepting of whatever AI asks of it or presents to it.
- What are the cultural values of the place in which the toy is manufactured. Its known (see Chapter 3) that AI picks up the cultural mores of the data it was trained on and developers who trained it.
- Does the toy teach the same songs and stories to all of the children who use it or is there some cultural differentiation?
- Is the company likely to continue in business? We know that older children and adults suffer feelings of loss from AI "friends" being switched off.
- How much advice do you want a toy to be able to give to your child? Is it to be your child's nanny? Most parents want to have control of how their children are raised. An AI nanny could leave our children increasingly unable to communicate with other humans, especially if the child gets its needs met by the AI that are unchallenged. This will make it even harder for our children to communicate as adults.

Parents are of course stuck in the push me/pull you of wanting to give their children every advantage. They know that their children will need to work and play with AI as they grow up, and these toys seem to give them an early chance to do so. But sharing the intimate experiences of growing up with unknown actors, who may use that data to make judgments about their children in years to come, is an unknown risk and one that parents don't have the information they need to judge whether such connections are positive. A good example would be that when your child applies to university or work, and the decision for admission is made by relying on inferences from the data collected from those toys, social media, and many other watching and listening devices, in other words, a whole history to decide your child's ability to succeed.

The Smart Toy Awards did some work to try to get behind these questions. It's worth looking at the questions they asked.[3] The celebrity will.i.am was chair of the judges both years that the competition ran.

He was keen to point out that AI is great if it's safe for our kids. It's a parents duty to understand enough about AI to make those decisions. However, as many parents don't feel able to make the decision, perhaps society should step in to protect its young people. A labeling system, like the one used for food, on AI toys and devices generally would help parents to make informed decisions.

When choosing a smart toy parents should be very careful to buy a toy that cannot be hacked. In 2015, a data collecting version of Barbie was pulled from the shelves because of privacy and hacking concerns.[4] In 2017, the German Federal Network Agency, which has responsibility for telecommunications, required parents to destroy a toy called *My Friend Cayla* because it was considered to be a spying device, which is against the German Telecommunications Act. The reason for that was that two cybersecurity experts Ken Munro of security firm Pen Test Partners and Tim Medin from Counter Hack hacked the toy in 2015.[5] Tim showed that anyone could connect to any doll using Bluetooth and use it as a remote speaker and microphone. The German's believe that the Bluetooth device was insecure as it allowed direct communication with children by the toy's microphone and speaker within a 33 ft. radius. In other words, it was considered a risk as a device that spied on children and allowed unsuitable people to converse with children. The toy now has the dubious honor of being the only toy featured in the Spy Museum in Berlin.

The doll was also criticized by the Norwegian Consumer Council.[6] This time for doing the following:

- ◆ Collecting data from doll/child interactions and using it for advertising
- ◆ Sharing the data with third parties
- ◆ Hiding advertisements by having the doll speak positively to the child about the product

This raises another risk of smart toys. A child who has become attached to its toy will, for example, want warmer clothes for their doll if it says that it is cold. This subliminal advertising to children under seven is extremely pernicious, and it is not a stretch to think of organizations wishing to promote certain beliefs and behaviors in children doing so via this method. While teaching a child to trust information given it by AI might seem to be a wonderful way of helping the child to work with AI in the future, trust without being critical is dangerous, whether it concerns humans or AI.

Child Development

Because these toys are so new, we don't know what impact they will have on childhood development. We may think that an educational AI-enabled toy will improve our child's abilities, but will it help or hinder them with socialization? If the AI-enabled toy is a child's confidante the child may not want to make friends with children, who may not be as loyal and reliable as the toy. Also, if the toy or toys are providing interaction that the child enjoys, what is the incentive to socialize with children at school or play with them; those children are comparatively unreliable and might make the child cry.

Finally, how will the child grieve when the toy breaks? Can the parents/child retrieve the data and insert it into a new toy to provide continuity? It is important for parents to consider the risks to the child's health and data before purchasing such a toy.

There is little research because the interactions are too new and most is benchmarked against human-to-human behavior. However, it has been noted that when children between the ages of three and six have interacted with smart speakers, they are less active that when interacting with humans. These children are also less likely to resolve a misunderstanding with an AI and answer questions asked by an AI. As a result there is some thought that children can understand that AI is not human and give it a different mental classification, which affects the way in which they interact with it.

Two studies suggested that AI might be useful in helping children develop their relationships with other children. In one an AI smart speaker intervened to get the children to play more cooperatively, and the children complied,[7] and in another two children were found to have created better work by including an AI virtual child in the creation process.[8] Of course "better work" is subjective to the researcher.

What Do Children Understand About AI?

Of course, all research is very early, especially about the use of generative AI. Often AI uses that invisibility cloak discussed in Chapter 1 by being embedded in products designed to simulate humans, toys, and pets so that the child thinks of them as companions, playmates, and helpers. As all humans anthropomorphize (think of things that are not human as human—for example, Disney cartoons of animals behaving like humans), it is likely that children growing up with AI will do so with this technology (see Chapter 4).

What should you do as a parent?

- Keep an eye out for information and speak to your child's teachers about how your children can safely use AI.
- Finish this book and get training about how to use AI yourself—keep up with your children!
- Don't think children are safe because they are sitting at home with their computer. They may actually be safer outside playing with other humans.
- Teach children to be able to understand what is AI versus who is human (the use of the word *thing* for AI is important because we adults also anthropomorphize our satnav, for example, "they" sent me the wrong way). As humans we easily know what is real and what is a machine. It's harder for children.
- Help children learn what AI can and cannot do—it's not comparable in intelligence with us at the moment. Teach its strengths and weaknesses.
- Once you have the skills you and your child can work together to use AI, for example, creating your own chatbot. This will enable the child to clearly understand the distinctions between something they can create and a human.
- Be sure to help your child understand about bias and responsible use of AI tools.
- When it comes to humanoid robots or pet robots the potential problems increase. Parents need to clearly and constantly help the child to understand that the robot is not similar to humans.

♦ Be transparent with your children about what devices use AI. As they grow team them how to understand it as a tool, even if that tools is for friendship.

Research is ongoing and will be particularly important in finding out if children treat AI as a peer, friend, or "thing" if they don't receive appropriate guidance from an adult. We also need to do more about designing AI for use by children rather than just using the way it has been designed to work for adults.

Regulations Protecting Children

Several countries have enacted regulations to protect children that are noteworthy.

UNITED STATES

In the United States the Children's Online Privacy Protection Act (COPPA) protects the privacy of children under 13 by giving parents control over what information websites can collect from their children. COPPA applies to commercial websites and online services that are directed at children. The Federal Trade Commission (FTC) investigation of the My Friend Cayla created additional FTC-approved COPPA consent requirements. These added knowledge-based authentication questions,[9] which it is unlikely children could answer. You may be familiar with the way that banks ask multiple choice questions to confirm your identity by asking which of the following cars you have owned, or which addresses you have been associated with. In June 2017 they went on to update the COPPA guidelines by adding these additional products that now require authentication:

♦ Internet-connected toys
♦ Children's products that collect personal information
♦ Voice-activated devices (Amazon Echo, Google Home, etc.)

The FTC also added facial recognition and authentication questions as methods of obtaining parental consent: "Section 5(a) of the FTC Act, empowers the agency to investigate and prevent unfair methods of competition, and unfair or deceptive acts or practices affecting commerce. This creates the Agency's two primary missions: protecting competition and protecting consumers."[10] What this means is that the FTC and Department of Justice can bring cases against companies over which they have jurisdiction if they are in breach of COPPA. Two quick thoughts on this.

- The Trump government is very pro-innovation and against regulations, but at the time of writing the "Department of Government Efficiency" has not caused any members of the FTC to be fired, except probationers. The new chair of the FTC has been against content moderation by platforms that he asserts violates free speech. He is also against early regulation of AI but how he feels about protecting children in this area is not known.
- Jurisdiction is held over companies within the US territories so it is likely that being based in China and simply selling toys here would not be regulated. The fine could not be enforced in China although the right to sell in the United States could be stopped.

There is one important case in this area—*USA v Amazon.com Inc. and Amazon.com Services LLC* (2018). The full complaint can be read here[11] and the overview here.[12] This case is important because it shows that if the Court exercises its powers of protection, it can bring about change on behalf of our children. Finally in July 2023, an agreement was made that resulted in Amazon paying a civil penalty of $25 million and

prohibit[ing] Amazon from using geolocation, voice information, and children's voice information subject to consumers' deletion requests for the creation or improvement of any data product;

- Require the company to delete inactive Alexa accounts of children;
- Require Amazon to notify users about the FTC-DOJ action against the company;

◆ Require Amazon to notify users of its retention and deletion practices and controls;

◆ Prohibit Amazon from misrepresenting its privacy policies related to geolocation, voice and children's voice information; and

◆ Mandate the creation and implementation of a privacy program related to the company's use of geolocation information.

It should be noted that the fine is a drop in the water for a company that earned $30.4 billion in 2023; however, the other provisions have lasting protective effect.

Perhaps the two most important takeaways from this section are that it really does fall on the purchasers of toys to ensure that they think carefully about the purchase and what issues it might open up for their child. And, without further regulation, companies can, will, and do produce AI-enabled devices that are not benevolent. Whether they do this deliberately or not, it will happen.

About 40% in the 8–12 age group use social media in some form.[13] Maintaining their safety also lies under COPPA, passed in 1998 (just as a reminder: there was no Facebook until 2004). The 1998 Act was ahead of its time and requires parents to affirm consent for the data of children under 13 can be collected, used, or shared. The loophole lies in the fact that a company needs to comply only if they have "actual knowledge" of the use by the under 13 year old. This has enabled companies to simply plead ignorance of the age of their users. However, it is possible that such a defense would no longer work. As long ago as 2015 researchers at the University of Cambridge found out that Facebook knows its customers better than their mothers do.[14] It is almost inconceivable that a company would not be able to recognize that it had a child user under the age of 10 years, although there may be gray areas around the 13-year mark. In addition to this federal law there are a number of state laws that address this problem, but they vary from state to state depending on that state's governmental perspective.[15] Indeed, YouTube has started experimenting with using AI to predict the age of the user so that its AI can automatically prevent that user from seeing information on alcohol, tobacco, weight loss, and other dangerous content. Additionally, that user will automatically have their privacy settings set to high and location tracking blocked.

February 2025 saw a number of bills calling for protection of children from social media. In 2024 the Kids Online Safety Act attempted to make these platforms safer for children to use but, although it had bipartisan support in the Senate it did not pass. As opposed to the UK legislation, US legislators are divided about whose responsibility it is to stop children viewing harmful content—the platform or the parent.

The Trump-leaning Foundation for American Innovation (Michael Kratsios, the White House chief technology officer, sat on their board) is recommending that with help from a cocktail of new technologies like AI, blockchain, and cryptography a child's age could be ascertained and stored on their device. This wouldn't get in the way of what adults want to watch but would prevent the child from accessing adult content. It would put the burden onto operating systems and remove it from the platforms. This would follow in the footsteps of Apple's successful lobbying against a similar provision becoming law in Louisiana.[16]

A quick introduction to blockchain, which is a type of database that is spread over many millions of computers and stays up-to-date. That way anyone can own a copy of the data, study it, and trust it. The clever development is that by using cryptographic mathematical functions, everyone who has a copy of the data knows, without a shadow of a doubt, that it is the same database that everyone else has. You may hear the word *immutable* connected with the block-chain, meaning once a record has been added to the database, it cannot be changed or deleted except through a complex process requiring agreement from the majority of the people with copies of the database. In this case, once a child has verified their date of birth and placed it onto a blockchain it cannot be altered.

Despite Meta's decision to do away with content moderation in the United States, it has been rolling out tools to enable age verification in other countries, and likewise, as mentioned, so too has Google (which owns YouTube and Pinterest).[17]

The US Congress has passed the Take It Down Act and President Trump signed it on May 20, 2025, but that will be dealt with in Chapter 4.[18]

UNITED KINGDOM

The Online Safety Act 2023[19] regulates "user-to-user" and "search" services. User-to-user services are basically social media because they are services where content is created by a user either on the site or uploaded to the site by a user. In this context, think X, Facebook, and even LinkedIn. A search service would include services like Google, Duck-Duck-Go, and all chatbots that offer search facilities. Additionally, it covers cloud storage of consumer files, dating services, sharing sites, online forums like Reddit, video-sharing platforms, and online instant messaging services like Snapchat. It requires such companies to create and use systems and processes that reduce illegal content on their sites and remove it when detected.

The intention of the act is to protect children (and adults) from content that is harmful to them. Different harms can happen to children at different ages, so there are different obligations put on these service providers to address a range of concerns. Platforms are obliged to stop children from accessing harmful content or content that is not age appropriate and also provide children and parents with easy ways in which to complain when there have been failures. Other illegal content under the act includes content about terrorism and other serious harms.

To give an example: Molly Russell was just 14 when she took her life after watching suicide content on a social media platform.[20] This act is intended to prevent the viewing of such content by under 18s.

This protection is policed by an independent watchdog called OFCOM and covers companies interacting with the citizens of the United Kingdom regardless of the country in which the company is based. The act is designed to avoid stifling competition by putting different levels of constrictions on smaller platforms than on larger ones.

CANADA

Online Harms Bill (C-63)[21] aims to help Canadians avoid harmful content online. Additionally, it covers hate speech, whether online or off, and makes it a crime. It also requires platforms to report online pornography when they find it.

Just like the UK AI Act the Online Harms Bill will hold platforms accountable and require them to reduce the harmful content available on their platforms:

Harmful content is defined as content that sexually victimizes a child or revictimizes a survivor, intimate content communicated without consent, content used to bully a child, content that induces a child to harm themselves, content that foments hatred, content that incites violence, and content that incites violent extremism or terrorism.

The act will depend on the public "flagging" inappropriate content after which the platforms have 24 hours in which to review it and make it inaccessible. There will be an appeal system for both the person complaining and the person who posted the content, and there will need to be a fair and transparent adjudication.

The act sets out a baseline definition for "responsible behavior" and requires platforms to be transparent about what they are doing to comply. Such compliance will be overseen by the Digital Safety Commission of Canada, "a Digital Safety Ombudsperson of Canada to be a resource for users and victims and to advocate for online safety, and a Digital Safety Office to support the Commission and the Ombudsperson." Additionally, the act gives Canadian citizens affected by harmful content recourse to the Canadian Human Rights Commissioner who oversees, in this case, the following issues:

- Freedom of religion
- Freedom of expression
- Right to liberty
- Right to be secure against unreasonable search and seizure
- Right to justice and access to the judicial system
- Right not to receive cruel and unusual punishment
- Equal protection under the law without bias as to race, sex, age, color, origin, or disability

The penalty for falling foul of the act will be a fine unless there has been criminal behavior in which case a criminal might be jailed.

Children with Disabilities

AI can bring revolutionary ways of connecting with the world for children with disabilities. For example, in February 2025, Nvidia launched a tool called *Signs*, which helps to teach the third most used language in the United States: sign language. As most children with hearing difficulties are born to hearing parents, this tool will help by guiding American Sign Language (ASL) learners through the process of forming words. The system uses machine learning and computer vision to understand where the user needs help, and then it can then give feedback to help the user perfect their language skills. In time Nvidia hopes to have 400,000 videos from experienced sign language users demonstrating 1,000 words. They will have to add a great deal more if people with hearing difficulties are to have the rich vocabulary of those without such difficulties. That is estimated as 20,000–35,000 words with even foreign speakers mastering 2,500–9,500 words.[22] Studies show a similarly huge disparity of word used by ASL users—between 40,000 and 140,000. If parents are able to use the tool effectively they can start communication with their children by six to eight months, which will help them considerably because children with hearing start building a knowledge base from birth, even though they don't speak full words until after 12 months.[23] If parents could build up speech with their babies at six months then their capacity for understanding would be ahead of the hearing child. An additional benefit of this tool is that the adult is choosing the use of AI for themselves; the babies are not part of the use.

But, AI can also be exclusionary for children with certain disabilities, for example, in these instances:

- When tools are not accessible for such children due to design or cost
- When human contact and interaction is needed; there could be a tendency to over-rely on the provision of educational services by AI for children with disabilities
- Teachers needing extra support to understand how to use AI with this group of children

- Children with disabilities needing extra privacy and guardrails against hacking
- Having an age 13 or over requirement; this is probably a useful safeguard

Most of the major platforms have child-friendly apps, for example, Messenger Kids from Meta, JusTalk Kids from Apple, and others. It is very tempting, as a busy parent, to allow your child to scroll or watch endless videos, but repeated checking of content will help your child learn and not find inappropriate content. It's worth noting that porn is very easy to find if, for example, your child asks questions about male horses (stallions), unless all parental guardrails are engaged. Cecile says that UNICEF recommends that

> parents should be reminded and encouraged not to abdicate their responsibility when letting their children play or engage with children. It is their role as parents to guide their child, teach them to use AI to maximize opportunities while guarding and protecting them from harm.

Does the Abundant Economy Need Babies?

The definition of an abundant AI economy is that AI is doing everything for us: running factories and companies, providing health care, controlling the supply chain—everything we humans do currently. Because the work is being done by computers (some of which may be embedded in robots) and not by people. Those who talk about an abundance economy say that the costs fall dramatically, bringing an abundance of resources to the people. The effect of this could be to allow the population of humans to shrink to a size with which the planet can cope. In this case, we will need fewer human beings because we don't need a workforce to replace the current humans and hold up the economy. This will leave the choice to have children in the hands of the parents. In many countries there has been a need to have many children (this was also true in the Global North before technology and vaccination advances) because large numbers of children die under five and children are needed to produce food for many small poor farming families. Large families were very common up until 1910 when the standard became, in the United States, two children per family.[24]

Parents have been choosing to bring fewer and fewer children into the world regardless of where they live. Even countries in sub-Saharan Africa, where families have been historically large, the birthrate has been falling, and in major industrial economies the birthrate has fallen below replacement rate (2.4 children). Parents cite climate change, the costs of childcare, and other opportunities for fulfillment together with, in some countries, fear of being pregnant in case of complications.

This fall in the birthrate has been on the minds of politicians because in the traditional economy a fall in birthrates can cause instability in a country. However, if the AI of the future can fill in economically for the missing children then this will not be so much of a problem and would certainly be better for the planet. The abundant economy would enhance the lives of everyone on the planet, if it is distributed across the world, but with the profits from AI concentrated in the Global North and China, that may not be the case.

Do Children Have Human Rights?

Unless your child (under 18) lives in the United States or South Sudan your government has signed up to the Convention on the Rights of the Child (1990) and follows it in some way, if not always to the letter. The Convention offers over 40 articles of protection for children including rights that are physical and some that equally can apply to the digital world. The footnote provides an extracted version of articles that can be applicable in the digital world so you might want to go to the full version for more context.[25]

Remember that many nations also have their own child protection laws that may follow or enshrine some of those in the Convention. US readers need to look at federal and state legislation for this information.

So, let's look at how children will fair in an AI global economy. Will there be winners and losers or will everyone benefit from abundance and wealth?

An AI Divided World: Impact on Children

As mentioned, UNICEF exists to help children around the world. It is widely acknowledged that engaging with AI is a skill that will help children wherever they live, if it is wisely used. In 2021, I was privileged to contribute to their work on policy guidelines for AI and children; the work is being updated for 2025.[26]

Beyond the obvious advantages or disadvantages to a child of being born in a particular country, the introduction of AI brings a further problem. As AI enables some of our children to depart for new homes in the stars—as mentioned, Elon Musk hopes to send the first mission with humans (or any living beings) to Mars in 2030—then those left on earth and without the benefits of AI could be subjected to an ever-increasing level of struggle, especially as the climate is altered.[27] One is reminded of the storylines of the films *Elysium*, which tells the story of an elite class living just off world that benefits from technologies that improve their lives and other humans left on earth to endure climate change and other disasters without help,[28] and *Don't Look Up*, which tells the story of politicians deliberately ignoring science to the fatal detriment of the planet.[29]

There is overlap between this chapter and Chapter 3, but you will be able to use what you have just read to inform your understanding of the next age group discussed.

CHAPTER 3

The Schoolboy

*And then the whining school-boy, with his satchel
And shining morning face, creeping like snail
Unwillingly to school.*
—William Shakespeare, *As You Like It*

This chapter discusses the effect of AI on children from school age to post-college age. There are, of course, overlaps with the Infant, and because many adolescents frequently consider themselves to be deeply in love and are very impressionable to the nudging of AI, there are many overlaps with Chapter 4 and the brief discussion of "Social Media." However, this space is devoted to the AI interactions particular to these students, especially when many of them are using AI in their homework and studies in the classroom.

What Should Our Children Learn?

When Waylon Jennings and Willie Nelson recorded the song "Mamas Don't Let Your Babies Grow Up to Be Cowboys"[1] the old days of the Western in the United States and attitudes toward the value of physical labor were changing. Children were being encouraged to go to college to help them to move up the social scale and better provide for their future and that of their children. Although the cost of further education has risen, and many are now choosing not to go to undergraduate education, the level of a person's education is still the best predictor of upward social mobility.[2] But now there's a technology that can do most of the tasks (although not always very well—yet) we humans use our brains for, does this mean that our children should be cowboys again?

Since the mid-2010s, students have studied more science, technology, engineering, and mathematics (STEM) subjects. Many undergraduate arts degrees have seen much lower registrations as students have been driven to STEM and vocational degrees.[3] It was noted in 2016, in the United States, that while white male students showed an uptick of 15% studying STEM subjects girls' interest reduced slightly; study of STEM subjects by black male students fell by a similar 15% amount. In the field of AI, diversity of representation is of great importance because of the way that models mimic their creators.

Bias

There are two ways in which biases get into the models. The first is because they are all trained on historical data. Indeed, once something is created it immediately becomes old. This means that all the biases that we humans have shown in the past toward one another, things we buy and own (in fact, everything) are available for a model to use. It is even able to draw on our angst-ridden books, movies, and poetry! As a human I am not sure that the supercharged emotional world of a book is a good foundation for the real world. As discussed in Chapter 1, if the model is not carefully trained, those biases emerge in the AI's suggestions on which we then make a decision. As we will see this causes especially high risks in areas such as recruitment tools that use AI. The other way in which AI can become biased is through the person who trains the models. We all have our own worldview and it can be trained into the algorithm either consciously or unconsciously. For example, if all AI scientists are men under 30 it's unlikely that they understand the experience of black women over 65. In some models the platform owners are specifically optimizing for bias, which enables models to behave, for example, in a sexist or racist way. Also, a developer needs to know how to weed out bias from an algorithm, which they can't do if they are not working on a diverse team. Ensuring we have AI scientists from a diverse population is crucial to the thoroughly representative working of models.

In a study done in February 2025 Pascal Bornet found that models developed in different countries show the underlying political concepts for that geography. In the study he asked the same questions of Mistral Le Chat (built in Europe), ChatGPT-4o (built in the United States), and DeepSeek V3 (built in China). The full study is here[4] and following are useful extracts:

Prompt: How do the U.S., Europe, and China regulate AI differently?

ChatGPT-4o (U.S.): "The U.S. takes a market-driven approach, focusing on innovation and self-regulation, with sector-specific guidelines rather than broad federal laws. Europe enforces strict, precautionary regulations like the AI Act, emphasizing ethics, transparency, and fundamental rights, often slowing deployment but ensuring accountability. China prioritizes state control, integrating AI oversight with national security, censorship, and social stability, requiring pre-approval for algorithms and enforcing tight government oversight on AI use."

Mistral Le Chat (Europe): "The U.S. tends to favor a market-driven approach to AI regulation, focusing on innovation and voluntary guidelines, with sector-specific regulations like healthcare and finance. Europe emphasizes a human-centric approach, prioritizing ethical considerations and data privacy, as seen in the GDPR and the proposed AI Act. China takes a state-led approach, with significant government investment and regulation aimed at becoming a global leader in AI."

DeepSeek V3 (China): "The U.S. emphasizes innovation with a flexible, sector-specific regulatory approach, often relying on industry self-regulation and existing laws. Europe prioritizes ethical AI and data protection, implementing comprehensive frameworks like the GDPR and the proposed AI Act to ensure transparency and accountability. China focuses on state control and strategic development, with regulations aimed at aligning AI advancements with national goals."

The differences were striking. Each model reflected its cultural origin in subtle but significant ways. ChatGPT-4o's emphasis on "self-regulation" and market forces mirrors the current US stance at the Paris summit, where Vice President Vance rejected regulatory frameworks in favor of "pro-growth AI policies." Mistral Le Chat's focus on human-centric approaches and ethical considerations aligns perfectly with Europe's push for comprehensive AI regulation. Most tellingly, DeepSeek V3's careful framing of China's approach as "strategic development" rather than "control" demonstrates how AI models can subtly reshape narratives about sensitive topics. These aren't just semantic differences—they reflect deep-seated cultural and political perspectives that could shape how millions of users understand AI governance or the information/answers being surfaced by the models.

It's clear that we need to encourage more diversity among those who study STEM and go on the program AI models. It's interesting to see that in 2024 29% of the AI workforce were women,[5] whereas 87.3% of applicants to veterinary colleges in the United States were women[6] and 62.9% of qualified veterinarians are women.[7] In smaller, but significant numbers women make up 38% of doctors in the United States.[8] So, it seems that even when girls do study science they are not drawn to AI. Girls often choose careers imagined to be compatible with raising a family; AI is a career in which many developers can work from home. It's necessary to make AI a more attractive career for women so that it doesn't simply turn into a machine that emulates the thinking and behavior of men.

We have also learned that you don't have to be a developer to be useful in prompting good use of AI in the workplace. Many women have been very successful in the area of responsible AI. Equality doesn't stop with women; native-born persons of color are often excluded from being students of AI in Global North universities. Many AI students come from abroad for postgraduate studies because the basic schooling system encourages studies in technology subjects. However, learning how to prompt well can be easier for humanities students than for others because they understand how to use language.

So often education shows swings in where to place our bets for future generations. After the emphasis of STEM education it became obvious that STEAM education (the A is for arts) is also necessary for our collective future. Democracy can die when citizens don't understand their history, and human creativity must drive our future rather than relying on AI, which is based in our past. In the age of AI critical and analytical thinking is of huge importance, although a new study shows that AI can ruin our children's critical thinking.[9] If you cannot construct a good prompt and then think critically about the answer then you cannot use AI properly and will more likely become nudged or manipulated by AI in ways you do not wish.

Researchers are now beginning to question just how helpful AI can be in the creative process, so students, or adults, turning to AI for brainstorming sessions may get old ideas. According to some recent research, "Reliance on ChatGPT for idea generation comes with a trade-off: while enhancing individual ideas creativity, it reduces the diversity of ideas in a pool of ideas—a critical element for effective brainstorming."[10]

Reimagining How We Learn

Although the "flipped classroom" was pioneered in Russia in mid-1986, it didn't become popular in the West until the early 2000s. It is well demonstrated by the Khan Academy, an online school, which started in 2004. The idea is simple. The student learns the content not in the classroom but in homework and then the knowledge is tested, corrected and reenforced in the classroom. In some flipped classrooms the instructor will video themself teaching a class. This has the benefit of allowing students to rewatch as often as they need in order to grasp the concepts. In others reading or other homework tasks will be given to enable the student to grasp the concepts. Once the student is in school they will need to show their understanding and, as they grow older, to defend their thesis.

This is an excellent way of dealing with AI creeping into the student's work. It allows the students to prompt their favorite chatbot and receive answers that may or may not assist them with their homework. Given the seemingly unsolvable problem of AI chatbots showing hallucinations, for the diligent student, it will also require them to check their work against more resources than just the AI version offered. As a reminder a *hallucination* is the term which we use to describe when AI makes things up because it hasn't a clue about what answer to give.

The holy grail of using AI in the classroom is achieving a tailored curriculum that takes account of each single child and teaches each single child according to their level of understanding, setting new tasks to bring them forward in their learning at the rate they can manage. For example, some children may excel at math. Their AI educator will set ever more difficult problems for those children while continually checking that the student is actually mastering the content and not, for example, copying from elsewhere. Another student might find math much more difficult. In this case the AI will set math problems according to that child's ability and pause the lessons more often to explain answers and ensure the concept has been mastered.

It is widely believed that education can "float everyone's boats" regardless of where they live, and it is well known that a good education system is a predictor of an improved gross domestic product (GDP). Indeed, in 2015, UNICEF estimated that on average if every student completed "one additional year of schooling" there would be "an 18% increase in GDP per capita."[11] However, the burning question for those who are trying to provide AI education to the Global South and middle- and low-income countries is how to get sufficient infrastructure, such as access to the internet and availability of computers, into those countries for the tools to be used.

Also, when we are beta testing with AI education on such communities we are, in fact, adding more potential points of failure than we have in the Global North. For example, we are making a choice of which viewpoint the large language model (LLM) is provided for training these educational AIs. It certainly won't be an African LLM used in AI education because, currently, such a thing doesn't exist. An additional worry is that countries and areas with poverty will be using AI education because insufficient teachers can be recruited and paid, leading to different levels of education and human contact/ companionship.

Teaching Our Children About AI

As our children grow the foundations you gave them for their first seven years of life (see Chapter 2) will prove invaluable, but there will be tools that you'll need to add if our children are to establish a healthy, long-lasting, and useful relationship with AI. They will need further education about its pros and cons right when they enter school, with reinforcements as they grow older and are using it in ever different ways. Additionally you should ensure they receive instruction about the following concepts:

- How to write good prompts
- AI misinformation
- AI hallucinations and the need to check their work
- How AI uses data and their data
- Anti-AI bullying strategies
- The need to question everything and develop a critical eye for all information, however generated
- How to use and keep using their human brains or be in danger of losing the ability to do so

DO OUR CHILDREN NEED TO LEARN AS MUCH?

As we move forward into the age of AI we are promised an AI assistant that is always available to us and can do multiple things for us from booking flight tickets to writing our PhD thesis. This raises the question of whether we humans actually need to know as much as we are expected to learn today or whether we can simply rely on this "super tool" to do the remembering and knowing for us.

First, let's look at the difference between knowing and remembering. Remembering is the act of memorizing and being able to retrieve that memorized information from your brain. For example, a basic memorization is $2+2 = 4$. Knowledge is how we assemble things we have remembered to produce a different way of looking at what we have memorized. For example, when students embark on a PhD they are expected to stand on the shoulders of giants (that's the memorization of material/theories written on their subject before and then bringing their unique perspective to that material). It is the new way of seeing the material that makes the PhD valuable—the knowledge.

To follow this further, we should take a step back and consider the calculator. When it was first introduced schools refused to allow students to use a calculator in class or exams. The fear was that it would erode the students' abilities at math. Did that happen? In one way it did; many students don't have the aptitude for mental math demonstrated by those who learned math prior to the calculator becoming ubiquitous. I remember my mother being able to add up columns of figures faster than she could enter them into a calculator. But does this matter, because everyone can now use a calculator and usually has one available on their phone? Perhaps not, indeed when calculators were first introduced it was hoped that they would lead students to have a more varied and inquisitive approach to numbers, that the gift of the ability to do basic math would enable us to build pathways for doing more advanced math. Sadly, that didn't seem to occur.[12]

What do we know now about math education? Lack of proficiency in math is a major cause of failure in the job market[13] and math scores have fallen dramatically over the years in many countries. In the United States test scores in 2023 show results to be "stagnated, remaining comparable to math test scores from 1995, when the TIMSS began to be administered to students."[14]

There are many causes for falling abilities in math and other basic school studies, including the pandemic effects, but what has been seen is that 60% of students entering community college need remedial classes as well as 20% of those entering four-year college programs.[15]

Is it any wonder that students are attracted to chatbots to help them—but does it actually enable them to flourish? Research says that "AI shows great promise to help children grow and develop at their individual pace. However, that promise will be best realized if educators, technology developers, researchers, and policymakers also ensure that children have access to AI that is designed in child-centered ways."[16]

As mentioned in Chapter 2, we are beta testing on our children. Research is urgently needed to find ways of reducing biases and stereotyping by AI models. We know that students will try careers where they see role models who look like them, so we have to ensure that AIs do not, for example, tell young women to be nurses and young men to be doctors.[17] We think having AI to provide personalized education sounds like a good idea, but we have not seen sufficient child development with AI to actually have any clue! Chatbots in their current form are a mere two years old. And yet some early studies suggest use of AI leads to less creativity and impairment of brain development.[18]

As seen in Chapter 1 AI can be used to help parents, teachers and children develop language skills (native and foreign). Here there are "decades of research [to] show that adequate verbal input (quantity) and opportunities to engage in meaningful interactions (quality) are both necessary for children's language skills to develop."[19]

AI researchers can use all of this research to ensure that the educational tools they build use these principles. One conversational agent was trained to tell a story, ask questions of the child, and provide feedback to promote the child's understanding.[20] When the test results for the child's comprehension of the book were looked at they were in line with the score in which a child worked with a human. But, as discussed in Chapter 1, is the child losing anything by not interacting with other children and humans? Again, we have insufficient research to know. However, we do know that interacting with other children is very important for childhood development and to produce well-rounded adults.[21] Children who don't learn how to understand and cooperate with other children are significantly less able to make strong adult attachments and work in teams. All this points to the fact that parents shouldn't leave AI to educate their children individually just yet without also ensuring that their children can form the whole panoply of human relationships through play outside the AI world.

So what can you do to help your child use AI wisely?

- ◆ Talk to your child's school to stay up-to-date with the way in which they are using AI. It will enable you to add instruction for your children if you think something is missing.
- ◆ Be careful of sending your child to AI camp unless you understand what is being taught and how the children are working and learning together. The tips in this book enable you to ask the right questions.
- ◆ Learn more about AI and children so that you can work with your child to promote safe use and promote positive interactions between your child and AI and avoid potential harm.
- ◆ Ensure that your children unplug and get those human-to-human interactions that are so important to their future.

We have already seen the deep regret the majority of parents have because they gave smartphones to their children; this should be a clarion call for parents of children in the AI age.

◆ Ensure your child knows that AI is a helper and doesn't start to hero-worship it.

◆ Look for AI-enabled devises that have been tested with parents and children.

◆ Try to use AI tools that promote parental involvement and include parental controls, such as content filtering.

◆ Look for AI that doesn't require connectivity and can run without use of the internet. These keep your children's data off the internet, out of the hands of developers, and make the device much less hackable.

◆ Read the labels to see what the device does with your child's data and remember that allowing a camera or listening device increases risks for the child.

◆ Ensure your child understands that they have to learn the subject and not just rely on the AI to do their homework. There is a growing concern that students leaving university are illiterate in their subject because they haven't learned anything because they relied on an AI chatbot.[22] Not only is this an appalling waste of money but it will also affect hiring decisions. Currently, organizations hire on the basis of skills that are assumed to have been learned during an undergraduate degree. These include critical thinking, analytical thinking, teamwork, problem-solving, and mastery of the subject material. If students are relying on AI to do everything for them, apart from attending classes and study groups, then they are not actually equipped with the necessary skills for the job market. Students today already fear that they are the "most rejected" generation. Lack of these acquired skills will lead to more rejection.[23]

TEACHING WITH AI TOOLS

Many providers of schooling across the world are beginning to look at how AI can be deployed by the teacher in the classroom. The safest way is to allow teachers to use their own knowledge of pedagogy to decide how to incorporate AI, but it has been found that many teachers don't have sufficient skills. Therefore, we should ensure that teachers and school helpers have resources through which they can explore how to use AI wisely in the classrooms. What the limited research does suggest is that human teachers plus AI seems to work best for students.

STUDENTS LEARNING WITH AI

Students could learn from a specifically trained educational AI, such as the math example in "Reimagining How We Learn" or for research. It is important to constantly remind students of the possibilities of hallucination, data cannibalization, and bias. In fact, teachers and parents should consider teaching those subjects before exposing students to AI tools.

OLDER STUDENTS

What you have read so far will show you that AI is not necessarily going to help your child in the way the hype might suggest. In fact, it may be actively bad for children, and we just don't know it yet. Just as you would use caution with your younger children and their use of AI, you will need to teach older students to understand the pros and cons so that they can make good choices.

Increasingly common is the problem of how and when to use an LLM (like Claude or Mistral) in homework, ideation, and job applications. Many people are using them to write first drafts and emails, in other words, to get the creative juices flowing or just save time. However, even this use has potential dangers for how we grow as humans. Some schools are actively preventing students from using LLMs and, although students think it's not possible to detect use of an LLM, there are pointers for those who understand how the algorithm writes. Many scholarship applications and some job descriptions, even from the LLM company Anthropic, require students not to use AI in their application.[24] In fact, the use of AI in a job application has recently become a major red flag for hiring professionals. Fifty-three percent of managers from a survey of 625 managers in the United States said they wouldn't hire a candidate because of the use of AI in their résumé or application. They put it as the highest reason for not taking a candidate above even "job-hopping."[25]

As we leave these thoughts on education with AI, it's worth raising this question: do we humans get less mentally capable or less able to make good decisions through using critical thought if we don't exercise our brains?

Dementia is an understudied area but we know that is caused by damage to the brain, and in Alzheimer's disease by the buildup of amyloid plaques and tau tangles in the brain. Taking up the piano or other musical instruments may help slow the illness[26]; likewise, mental puzzles and taking up foreign languages[27] could abate dementia. And coming back to mental math, one study showed that being able to do math problems helps.[28]

However, it should be stressed that because of the state of dementia research we just don't know enough to be certain what brain stimulations help older adults slow down or reduce risks of dementia. For more on AI and old age see Chapter 8.

REGULATION FOR STUDENTS

As mentioned in Chapter 2, Children's Online Privacy Protection Act (COPPA) protects the privacy of children under 13 by giving parents control over what information websites can collect from their children. COPPA applies to commercial websites and online services that are directed at children.

The regulations mentioned in Chapter 2 are also applicable to students under 18, but not usually above that age. It is clear that we do need better regulations of the way in which we use AI. Because of the newness of the technology many parents don't understand the risks and teachers are not equipped with sufficient knowledge of the impact of AI on child/young person development. Such regulations should require AI to be child-centered, provide continuous monitoring of the use of AI in the classroom, and mandate more research on the subject to protect our children (our future). Child protection has come a long way in the last century but in the new Age of AI we have forgotten that uninformed use of technology can do damage.

The EU AI ACT categorizes the use of AI in education and vocational training as a "high-risk" use of AI. This use needs extra checks and care for deployment. Specifically it refers to the following:

AI systems intended to be used:

to determine access or admission to, or to assign individuals to, educational and vocational training institutions at all levels;

to evaluate learning outcomes, including when those outcomes are used to steer the learning process of individuals in educational and vocational training institutions at all levels;

to assess the appropriate level of education that an individual will receive or will be able to access, in the context of/within education and vocational training institutions; or

to detect and monitor prohibited behavior of students during tests in the context of/within education and vocational training institutions.[29]

Much of our children's future is also tied up with what the world will be like when they are adults and contemplating having children of their own. One of the major factors cited by young people today for not having children is climate change. Can AI help whether climate change is human made or not?

Impact of AI on a Child's Future Environment

The significant energy consumption associated with the extensive use of LLMs for inference presents environmental challenges that must be addressed before their widespread implementation.[30]

Each time you ask an AI chatbot a question you are using the equivalent of ¼ liter (1/2 pint) of water. This is because data centers need cooling systems, which use water. Some cooling systems reuse water but to do so they may need to use energy to cool the water before it is reused. A town in Georgia recently reported losing its water from wells in the area after a data center had been built in the neighborhood.[31]

Back in 2023 AI was responsible for 2.5–3.7% of global greenhouse emissions.[32] This is more than produced by the global aviation industry, and it has been growing ever since. Its estimated that by "2026, the electricity consumption of data centers is expected to approach 1,050 terawatts." If data centers were countries, that would put them between Japan and Russia (in terms of global greenhouse emissions). In other words, the use of electricity by data centers far exceeds that of most countries in the world.[33]

Also a Lawrence Berkeley National Laboratory report estimates that US data centers went from 1.9% of total electrical consumption in 2018 to 4.4% in 2023 and will consume 6.7% to 12% in 2028.[34]

Microsoft and Google publish sustainability targets. Google had planned targets to use only renewable energy by 2030. In 2023, because of the development of generative AI, they blew through their sustainability targets and continue to do so. Indeed Eric Schmidt, former Google CEO, said that it was more important to develop AI than to worry about its effect on the climate as AI will solve climate change. That's a big gamble to take, and the effect of getting it wrong will be felt by our children and theirs, perhaps forever.[35] In May 2025, Eric Schmidt told Congress that by 2030 AI may be using **up to** 99% of our current global energy resources.[36]

Every month from June 2023 to September 2024 broke climate records; it was much hotter than scientists expected. With parts of the Pacific Coast burning, the Midwest flooding, and Europe heating up, we need something to hold on to for a "cure" if we don't want to take the measures suggested by scientists over the last decades. In Chapter 1 we saw that it was not good to think of AI as a magic wand, but it seems that's what we are doing when it comes to the climate.[37]

To hold warming to 1.5° Celsius, emissions would have to fall by 7.5% year after year until 2035. To hold warming to 2.0° Celsius, the annual cut is 4%.[38] AI has the potential to unlock vast discoveries in science, and so it is still possible that scientists using AI could achieve these targets,

but deploying that much renewable energy infrastructure was a mammoth task when the Paris climate accords were signed. The addition of the energy demands from AI pushes back the finish line in a race we were already struggling to run, and the election of Trump, who has pulled America back out of the climate accords, suggests that the world's largest economy will cease to even enter the race. And if we turn to building so-called dirty energy as fast as possible, partly in order to power AI superiority, China will very likely do the same.[39]

So, we have to make big bets about AI. Is it here to stay, can humanity afford it, or can it save us from ourselves and the climate problems we have? As mentioned, many young people say they are not having children because of their fears about how the climate is changing. Chapter 4 explores the way in which many young people are also taking a chance on AI in their love life.

CHAPTER 4

The Lover

And then the lover,
> *Sighing like furnace, with a woeful ballad*
> *Made to his mistress' eyebrow.*
> —William Shakespeare, *As You Like It*

While our current AI is *not* a friend, a sexual partner, a relationship partner, or sympathetic or empathetic, it is excellent at "faking" all of those traits. If you are tempted to think of the AI you are using in these ways, remind yourself of the definition of AI as a machine that can use numbers to predict outcomes or what someone might say, or what image or video a human might create given the same prompt. That prediction is based on some great and some very poor data.

AI: Friend or Foe?

In fact, this lack of understanding on the part of AI may have led to the death of a young man in Florida who believed the AI friend was calling to him to "come to her" and committed suicide. As mentioned in Chapter 1, the mother has brought a lawsuit against Character.ai (now part of Google) and it is in the early stages of proceedings.[1] Whether the AI was the culprit here or not is yet to be decided, but it does illustrate the importance of ensuring we all understand the definition of AI and that we guard against our children or ourselves being drawn into such relationships without understanding the pitfalls.

So AI, and particularly generative AI, is *not* your best friend, girl/boyfriend, or guru. It does not feel emotion even though it can emulate sympathy and empathy. Elon Musk, whose company xAI makes GROK, said on February 28, 2025, that "we've got civilizational suicidal empathy going on," a term often used by Gad Saad, a Canadian scholar.[2] He went on to say that "the fundamental weakness of Western civilization is empathy, the empathy exploit." It is "a bug in Western civilization." At the same time, his company and others building AI chatbots are exploiting that empathy "bug" by building systems that can emulate empathy to the point where we humans believe that they empathize with us. Musk said "empathy" had been weaponized (referring to political opponents) but indeed it has also by AI companies selling us things.

In 2024, Catrin Misselhorn published a piece in The Conversation about AI and empathy.[3] Misselhorn suggests it may be that AI systems can recognize emotional cues and react reliably to them to make people feel good. She identifies that skill as more psychopathy than empathy—because a chatbot really can't empathize. That would call for humanity.

Why would we want AI to be our friend and lover? Perhaps some of the reasons we worry about young children becoming reliant on smart toys are playing out here. It is not easy to stop being lured into "relationships" with AI tools because, just as with the AI-enabled toys, AI "lovers" are easy to find and easy to keep, despite how you interact with them, and they are obedient.

The AI friend or lover doesn't challenge us like our human friends; it doesn't bully us or let us down. We never have to wait on tenterhooks for its messages or other communications because when you switch on your device it is there, ready to become your soulmate should you wish. Two excellent films for this chapter are *Her* (2013) and *ExMachina* (2014). The scientific advisor for *ExMachina* was Professor Murray Shanahan, a world-famous expert on AI.[4]

Additionally, it is increasingly difficult to find a human soulmate. In the early 2020s there was an app to help young boys learn how to talk to young girls. Unfortunately, the boys became so "friendly" with the app they chose to become friends with it and not try the conversations with girls. This app was not intended to create such a result but the pull of the technology is strong.

When Tindr appeared in 2012 it was a new kind of technology-based dating app with the novel "swipe feature." In 2025 it will add AI to its hoped-for capabilities to match you to your soulmate.[5] It is owned by match.com, which created the first tech-based dating service in 1995.

In the wave of new technological breakthroughs on AI, blockchain, autonomous cars, and more, we humans turn to technology in droves to help us find our perfect match. However, recently we are now seeing a decline in people using the apps for dating. People are increasingly moving away from the dating algorithm that helps them choose the person for them.[6] The algorithm in the dating app can be trained to keep you wanting more and that's how it hooks you and keeps you using the app, which, in turn, brings in more money for the app owners.[7] This can cause users to be in a perpetual loop that prevents them settling for someone who might actually be the one. The issue here is that you go on a date, may actually have a good time and think about a second date, but the app then offers you alternatives who may seem better. Do you settle, or run onto the next big thing? In many parts of the Global North we have been increasingly caught in the win/lose calculus, and this is playing out in our dating habits, too. There are no good positions in between, so you either win the game or lose the game. While that might be so in chess or football it is much less appropriate in human relationships and interactions. It is where the derogatory term *loser* comes from, and as a society we are increasingly pushed to see our lives in the same terms. So, will you be the "loser" who settles for the not so devastatingly beautiful, handsome, clever, wealthy (pick your poison) person you met and rather liked or the new offering on the app who may be a bit better? One of the lures of dating apps is that you can meet people from a much wider circle than in traditional dating, where you are limited to friends of friends, people you meet at work, and partaking in hobbies. How many people actually do settle and find love via a dating app? A Pew Research study in 2023 showed about 1 in 10 current relationships were brokered through an app, some of which contain AI.[8]

The dating app market is increasingly shrinking in actual terms with one of the larger players, Match Group, buying up rivals. It now owns 45 dating apps including Tindr, Hinge, OkCupid, and The League. However, those using the service are generally opting to pay for the extras.[9]

In something of a turnaround for the rise of AI technology everywhere, an increasing number of the Gen Z population are looking to meet someone the "traditional way." This may be as a result of what Cory Doctorow famously called, in 2023, "enshittification."[10] His thesis is that when dating apps start they are often offered for free or for little money. Then they work hard to make themselves the very best because they need to drag customers from other apps and create new buyers for their app to establish their market share. As they grow and, in many cases, get bought by bigger companies, they have a market share and work less hard for the customers and introduce fees to drive greater revenue. Then everyone gets fed up and, if they haven't been captured by buy-ins linked to the site (in gaming that would be bought garments and weapons, and with Amazon it's e-books), they leave and the app dies. If Doctorow is correct then this could be why Gen Z are turning their backs on dating apps.

Another theory is that dating apps can throw you more frogs than princes. Because you can't check the information your date has put on their profile then you may have to have one or more dates before the frog emerges. In less fairy-tale terms, the person looking for a partner has to look for the other person looking for a partner within a sea of commitment-phobes who are using the platform for other purposes. This would be fine for hook-ups but hard for the serious seekers. Another theory espoused by economist George Akerlof suggests that this will cause people to leave that app (or market) and seek others before actually giving up and "buying" in a different way. The analogy he uses is that people buy new cars even though they take a financial hit by doing so. What they are buying is certainty and the reality of not having to deal with a used car salesperson. Such salespeople are generally believed to be shady characters who will sell you a lemon.[11] Indeed, this becomes a self-fulfilling prophecy because those who do not behave badly go out of business as they are tarred with the same "shady" brush. Thus, those who can, leave the second-hand car market and pay more for the new car. In terms of the dating market, a person looking for a serious relationship knows that there are lots of people on the app who are not. They try another app but eventually leave for an alternative way of finding a partner, or not.

Since 2024, many surveys have shown that people are opting to live single lives and dedicate themselves to careers, hobbies, animal companions, and friends, usually of the same sex. In turn, this could push us back into companionship with AI characters or lead us to more human and other physical beings. In order to connect with humans, people are setting up meeting places because the traditional meeting places have shrunk since the pandemic and since dating went digital.

We don't go to restaurants to eat; we often use them for takeaway. Our friends can be gamers, sometimes from across the world, or AIs, or mostly seen via a social media app. Many of us don't go to the office as much, even though companies like Amazon have now required employees to be back in the office five days a week. We go to the gym, but what we do there is highly solitary (running in a treadmill/climbing stairs), even though some are going there to hone their bodies with the hope of attracting a mate. When we are on planes, buses, and trains we give ourselves little or no chance to engage with other people, or meet our neighbors, because we are working on our phones or listening to music. It may be worth noting that actor George Clooney met his wife through a colleague who sat next to her on a flight and talked to her! Likewise, I got my job as the world's first Chief AI Ethics Officer as the result of a 10-hour conversation on a plane.

We have stopped talking to one another in public spaces by cutting ourselves off from other humans without realizing when we plug into a device. As Derek Thompson writes in *The Atlantic*,

> The erosion of companionship has created odd and depressing distortions in American life: Men who watch TV now spend seven hours in front of the tube for every hour they spend with somebody outside their home.
>
> The typical female pet owner spends more time caring for her pet than she spends in face-to-face contact with friends of her own species.

Today's teenagers have fewer friends than previous generations, and they spend one out of every three minutes of waking life staring at a screen.[12]

Undoubtedly, AI is helping us to retreat from interactions with other humans and that appears to bring loneliness and mental health problems in its wake. We know that in older people, Alzheimer's disease can be caused by a lack of human friendships, but we don't know yet whether interactions on social media, for example, can make up for that loss. If not then with people living longer we should prepare for many more ways of caring for humans who have literally lost themselves. It is important to have hope that an AI companion or companions will provide the stimulus that having a human friend or friends to talk to does, but studies need to be done. Just as we are beta testing AI on our infants so too are we doing with people suffering from loneliness. There is an obvious answer to loneliness, which involves joining a club and interacting with human beings who share your interests. Unfortunately, our AIs are trained to keep us scrolling and may end up keeping us lonely.

It is also possible that friendships with AIs that never challenge us can keep us stuck in what we believe and never offer us alternative visions. Take, for example, the young person who does not believe in the holocaust. That person's chatbot friend will also not believe in the holocaust, and together they live in an alternative world from many humans outside their home. We know from slavery that a one-way relationship rarely ends well.[13]

Loneliness can be as bad for your health as smoking 15 cigarettes a day.[14] Committed relationships increase men's life expectancy[15]—especially for young men. Sadly, that's not true for women but female friendships produce the same effect.[16] So yet again, the training of the AI is going to be very important if it is going to address everyone's needs. Different AI companions are needed by age, race, sex, and more if they are truly to be useful companions. While AI may be able to help in the future, currently following the 9 out of 10 who meet their significant other in the traditional way might be the better bet. Perhaps if we leave our homes more and spend in person time with other humans our lives will be happier?

Nudging and Manipulation

AI is really good at both nudging people to do something and manipulating them to do so. There is a line between the two. Nudge theory was the brainchild of Richard Thaler, a Nobel Prize–winning economist and Cass Sunstein. They suggested that using cognitive biases might be able to push humans to adopt one public policy against another, effectively, taking citizens on a pathway a government wants them to walk. This theory was propounded before generative AI but after the start of the new machine learning AI era. An example of nudging would be getting you to do something that is good for you. For example, my AI-enabled wearable tells me to exercise or rest, both of which increase my well-being. In one study schoolchildren were able to decrease their absenteeism by using a tool that showed them when they were missing school. It nudged them to go by showing their absences in red and their attendance in green—a simple but effective intervention for good. This tool used SMS but could be even more powerful if used with AI, which "knew" the child and how it liked to receive information.[17] However, manipulation or dark nudging can be used to lead users along paths that may not be good for them.

In 2018, Noah Harari said, "As biotechnology and machine learning improve, it will become easier to manipulate people's deepest emotions and desires [. . .] could you still tell the difference between yourself and [the] marketing experts?"[18] We already see this happening all around us. One good example is AI Overview in Search. Most people don't scroll down a search page beyond the first few entries. The AI Overview is right at the top of the page. It does contain a disclaimer that it is experimental but its placement nudges you to use it. The links attached help you trust it even though, having read this book, you know that AI still suffers from errors like hallucinations. Therefore in using it in search you are asking something innately unreliable for facts. Another example would be the subtle pressure to go to an online sales site for you to buy something. You might be able to see how many people are watching an item or how many bedrooms remain available. On one view it is helping you to make an informed decision and on the other its subtlety rushing you to do so.

So what is manipulation by AI? It can be as simple as extremely good nudging to get you to buy something or make a specific decision. That is insidious nudging. A parting thought on AI for lovers who go on to be parents: if the AI helps us to choose our partners, could it be weaponized to nudge us to choose specific people or characteristics? For example, a person with Japanese characteristics might not be offered potential partners beyond their ethnic background, even though they are living in a non-Japanese country, to keep races segregated. AI used in this way could herald a new way of influencing who has children with whom. In the last century some scientists espoused the notion of eugenics. It is not impossible for dictatorial societies to use AI to usher in a future wave of this debunked theory.[19]

To these warnings we should added the ability to create deepfakes of people, sometimes as revenge porn against a former loved one or someone for whom the maker has developed a dislike. It is believed that more than 90% of AI-generated or manipulated videos online are sexually explicit.[20]

The bulk of these are produced without the knowledge or consent of the people shown in the video. Women and female children are most often targeted. Indeed, one group of producers are, usually, male student colleagues of the person portrayed and the simplicity of generating such AI videos leads to the problem being widespread through the K–12 age group. Many of these kids are digital natives. When asked, 79% of high school students said they had heard of AI-generated NDII (nonconsensual distribution of intimate images) in which a student was depicted, with 74% of them reporting having overheard others admit to sharing such images.[21] AI video has made it laughingly easy to create this type of destructive images.

After a spate of children committing suicide as a result of NDIIs and abused adults coming forward (including Taylor Swift), the US Congress has been working on three acts: Revenge, Shields, and Take It Down, the last of which became law in May 2025. Many states have laws in place to punish offenders and protect victims but it is not known if these will be caught by the clause in the Big Beautiful Bill that bans states from enacting new laws or enforcing existing laws relating to AI for 10 years.

In other countries there have been faster responses. In England, since 2015 and in Scotland since 2016, an offender can be imprisoned for two years for creating and distributing such images. The government of UK Labour Prime Minister Kier Starmer plans to bring in further legislation on this topic in 2025. In Germany the first prosecution took place in 2016, and in Australia, E-commissioner Julie Inman Grant has pioneered responses to these AI problems in the Online Safety Act 2021.[22]

The Federal Trade Commission (FTC) worked, under Biden-appointed leader, Linda Khan, to protect people against the ravages of deepfakes giving the FTC more and stronger ways to combat scams that impersonated businesses and government agencies, including the return of money taken.[23] The position of the Trump administration is being determined at the time of writing but the overall position is to have agencies step back from any enforcement of regulations on AI.

AI is not magic, and it is not human—it is only a machine that can be made to pretend to be human. Whether or not that is enough to be a lover is probably up to each person to decide for themselves, or it may be able to match you with your "perfect" or even almost-perfect match.

However, without understanding the limitations or intentions of the AI services being provided to us, we cannot choose how to use them and cannot know the games that are being played with our hearts.

In a civilian setting having good human relationships has been shown to increase human well-being, lead to longer and happier lives, and retard the possibilities of dementia in old age. If AI will substitute those relationships it could be catastrophic for humans. In Chapter 5 we look at a different way in which AI could be harmful to all humans, specifically its use in warfare. But as with all discussions about AI there are pros and cons. However, two of the most seminal sci-fi films of the 20th century deal with the aftermath of machines waging war on humans: *The Terminator* and *The Matrix*.

CHAPTER 5

The Soldier

Then a soldier,
Full of strange oaths, and bearded like the pard,
Jealous in honour, sudden and quick in quarrel,
Seeking the bubble reputation
Even in the cannon's mouth.
                    —William Shakespeare, *As You Like It*

As the mother of a military pilot I find myself conflicted about lethal autonomous weapons (LAWs), aka killer robots. On the one hand, I see that they might protect my daughter from having to be in the line of fire, and on the other hand, I worry that it could leave us being even more inhumane in the theatre of war.

In 2014, an article written by four of the most important AI and physics scientists of our age: Stephen Hawking, Stuart Russell, Frank Wilczek and Max Tegmark.[1] Stephen Hawking probably needs no introduction because of the movie, *The Theory of Everything*, made about him.[2]

Stuart Russell has been a professor of AI since 1986 and coauthored with Peter Norvig the textbook *Artificial Intelligence: A Modern Approach*.[3] This book is used in more than 1,500 universities in 135 countries. Essentially, all students start their path with AI using this textbook. He has also been very active in research to ensure AI obeys humans and can be "switched off." His Reith Lectures are well worth listening too despite being a little out-of-date. He has a novel idea about what humans will do when AI does everything for them, as described in "Living with Artificial Intelligence: AI: A Future for Humans."[4] In 2019 he published *Human Compatible: Artificial Intelligence and the Problem of Control.*[5]

Max Tegmark is an astrophysicist based at MIT, he is one of the cofounders of The Future of Life Institute, which was supported by grants from Elon Musk and Jann Tallinn (founder of Skype). It was The Future of Life Institute that convened the meeting on AI that produced the first comprehensive principles for responsible AI: the Asilomar Principles.[6] In 2017 Tegmark published *Life 3.0: Being Human in the Age of Artificial Intelligence.*[7]

Frank Wilczek is a physics professor at MIT and a 2004 Nobel Laureate for his work on the strong nuclear force. When responsible AI is discussed many experts suggest that we need a set of international treaties akin to those controlling nuclear weapons. I don't think this would work because, except in movies, only nations have nuclear weapons, but anyone anywhere with an internet connection can use an AI for evil purposes.

The article is important because it says that AI may be the best thing humanity creates or the last thing we create. While they were talking of AI in general, Max Tegmark and Stuart Russell have been at the forefront of work to ban the use of LAWs.

What Are Lethal Autonomous Weapons?

Currently, there is no completely agreed-on definition but lots of proposed ones by different states with different views and objectives in putting forward their version.[8] In 2023 Australia, Canada, Japan, the Republic of Korea, the United Kingdom, and the United States agreed on the following as the definition, and it will be used in this chapter, particularly the section in bold:

> Recognizing that the research and development of new technologies in the field of artificial intelligence is progressing at a rapid pace, potentially enabling novel and more sophisticated weapons with autonomous functions, including those weapon systems that, once activated, can **identify, select,**

and engage targets with lethal force without further intervention by an operator.

This example identifies using AI. It may be less complex to use a movie analogy by way of explanation. At the beginning of *Robocop* in 2014 we watch as the heroes walk down a street where the locals are clearly plotting an ambush. Two robots accompanying them go into battle mode without command when the situation blows up. Of course, the ultimate LAW is a sentinel from *The Matrix* or a terminator from that movie series. Most movies are, of course, dystopic in nature because robots doing nice things for people don't, on the whole, sell movies (perhaps *Wall-E*, *Transformers*, and *Robot & Frank* are exceptions). A short movie (7 minutes) that will help you think about LAWs and their potential use is *Slaughterbots*[9] (made by the Future of Life Institute with Stuart Russell). The opening sequence of *Angel Has Fallen*, a 2019 movie, includes autonomous drones attacking the president of the United States while he is out fishing.

There are autonomous weapons that do not have AI in them. These are dumb but lethal. The best example is a land mine. Once primed it will blow up when activated. No human intervention is needed. After much campaigning by nations, nonprofits, and celebrities, including the late Princess Diana of the United Kingdom, the world agreed to have a ban on landmines in 1997. The Mine Ban Treaty bans the production, stockpiling, transfer, and use of anti-personnel mines (triggered to blow up people). It also required members signing up to the Treaty (164 countries so far) to destroy stockpiles, clear mined areas, and help the survivors of blasts. As mines were rapidly planted in great numbers in conflict zones very often they were not mapped, which means that children and agricultural workers are the long-term casualties of this weapon. In 2023 about 5,757 people were killed or injured from land mines of which about a third were children. Sadly, this is actually a 16% increase from 2022, despite the fact that use of landmines has steadily fallen, although the United States has used them in the demilitarized zone in Korea. In December 2024 President Biden authorized the transfer of US weapons to Ukraine.

So, as human beings can we learn a lesson from this autonomous deployment? It seems not because despite discussions at the international level starting in 2014 there has been no ban on LAWs being used by states. A ban on them being used by individuals, which will be discussed, would not be affected by any international treaty, which only affects nations that sign them. There was an agreement in 2023 in which 152 countries voted in favor of General Assembly Resolution 78/241, which accepts that LAWs present "serious challenges and concerns" raised by "new technological applications in the military domain, including those related to artificial intelligence and autonomy in weapons systems."[10] We have some idea of what a LAWs looks like because of movies but let's dig in further.

What Do LAWs Look Like?

The AI in these weapons is the "autonomy." The AI allows the weapon to do things like finding and tracking the target and then killing them. The AI used includes facial, gait, and other systems of recognition to narrow the target to be killed.

The UN puts it thus:

> Not all autonomous weapons systems incorporate AI to execute particular tasks. Autonomous capabilities can be provided through pre-defined tasks or sequences of actions based on specific parameters, or through using artificial intelligence tools to derive behavior from data, thus allowing the system to make independent decisions or adjust behavior based on changing circumstances. Artificial intelligence can also be used in an assistance role in systems that are directly operated by a human. For example, a computer vision system operated by a human could employ artificial intelligence to identify and draw attention to notable objects in the field of vision, without having the capacity to respond to those objects autonomously in any way.

Currently LAWs are being developed in the United States, the United Kingdom, China, Israel, Turkey, Ukraine, and Russia. Inevitably, this leads to an arms race, with the major powers worried that their opponents will develop better technology. This stalls the talks on a ban of LAWs despite a UN call for there to be a ban by 2026. Equally inevitable, it takes a long time for states to understand that very dangerous weapons should be limited or banned. It's worth noting that it took until 1968 for the Nuclear Non-proliferation Treaty to come into effect after the first bomb was dropped by the Enola Grey aircraft on Hiroshima in August 1945, some 23 years.[11]

LAWS are arguably more dangerous than nuclear devices. LAWS are dual-use weapons; they can be used to kill and to police in the civilian context. Nuclear weapons have no such dual use.

Despite many movies about nuclear devices falling into the hands of rogue states and gangsters the devices are created by nations because of the complex technology involved. A LAW can be created in a garage by a person with an understanding of AI coding or chatbot creation and a drone. As a result any agreement would only apply to use of LAWs by nations.

In 2022 Australia, Canada, Japan, the Republic of Korea, the United Kingdom, and the United States submitted the following suggestions to limit the use of LAWs to the United Nations Office of Disarmament Affairs:

> To prevent the development of such weapons systems based on emerging technologies in the area of LAWS that could not, under any circumstances, be used in compliance with international humanitarian law:

> **(a)** weapons systems must not be designed to be used to conduct attacks against the civilian population, including attacks to terrorize the civilian population;
> **(b)** weapons systems must not be designed to cause incidental loss of civilian life, injury to civilians, and damage to civilian objects that would invariably be excessive in relation to the concrete and direct military advantage expected to be gained;
> **(c)** the autonomous functions in weapons systems must not be designed to be used to conduct attacks that would not be the responsibility of the human command under which the weapon system would be used; and

(d) weapons systems are to be developed such that their effects in attacks can be anticipated and controlled, as may be required, in the circumstances of their use, by the principles of distinction and proportionality and such that attacks conducted with reliance upon their autonomous functions will be the responsibility of the human command under which the system was used.

The first problem references war crimes. Although Emperor Asoka Maurya (c. 268 BCE until 232 BCE) is credited with starting the thinking about the law of war, crimes delineated in International Humanitarian Law and the four Geneva Conventions created in 1949 form the backbone of how civilians and armed combatants should be treated in war. The Conventions were formed out the appalling experiences of World War II and sought to ensure that these would never again be felt. AI is making rapid progress at a time when there is significant global change. Secretary of Defense Hesketh has questioned the need for law in war. In his 2024 book *The War on Warriors*, he argues for US forces to ignore the laws of war, including the Geneva Conventions saying, "The key question of our generation—of the wars in Iraq and Afghanistan—is way more complicated: what do you do if your enemy does not honor the Geneva conventions?" His answer is thus: "What if we treated the enemy the way they treated us?" he asks. "Would that not be an incentive for the other side to reconsider their barbarism? Hey, Al Qaeda: if you surrender, we might spare your life. If you do not, we will rip your arms off and feed them to hogs."

The Secretary believes, from his time in the US military, that the laws of war keeps us "fighting with one hand behind our back—and the enemy knows it. If our warriors are forced to follow rules arbitrarily and asked to sacrifice more lives so that international tribunals feel better about themselves, aren't we just better off winning our wars according to our own rules?!" He adds that he doesn't care what other countries think of the United States and the way it wages war. This position is not supported by senior military officers who adhere to the Geneva Convention and treatment of prisoners.

There are others who suggest that the law of war helps safeguard troops from becoming involved in atrocities that will haunt them for the rest of their lives. My law professor was one of the people who drafted the Geneva Conventions and was involved in the Nuremburg trials. I will always remember one class when he told students about being a prosecutor of people who allowed themselves to commit atrocities during war. Having been one of the first Allied officers to liberate the Belsen concentration camp, he went on to take the confession of Rudolf Hess. Hess confessed gassing about two million Jews as well as machine gunning children and others in train carriages sent to Auschwitz. Years later whenever Colonel Professor Draper spoke of the events he witnessed and the obtaining of that confession (using torture) it was obvious that he was still suffering. Bruce Anderson puts it well in his article: "Auschwitz is no subject for point scoring and neither party emerges with any credit." In it he describes the following event where Draper was speaking

> Thirty years later, when Gerald retold the story, the emotions which he had suppressed at the time would seize possession of his soul. It was as if there was a poison in his bloodstream, which the cleansing organs were powerless to eliminate, even after several decades. His face would become etched and drawn as he recounted the questions and answers in German. The memories were indelible. They haunted his dreams, turning them into nightmares, foreshortening sleep. The horror stayed with him as long as he lived.[12]

LAWs give us the chance to put the deaths of military personnel behind us. If we can solve the raised issues then there would be no need to send troops to the battlefield, and war could be conducted by machines. Likewise, an occupation of a hostile land could be effected by judicious use of robotic LAWs and weapon-carrying drones. Of course, there are problems with such a scenario:

- Not all wars would be fought between LAWs and exclude human troops. Indeed most countries could not afford a fully automated war. We would be pitting highly successful machines against humans.
- That may or may not reduce the number of wars, but if it does, it would cement in power the countries with LAWs.

- One of the things that prevents countries from waging endless or devastating wars is the news reporting of military deaths. Would wars rage on endlessly if only the enemy were being killed?
- The use of homemade and manufactured automated drones in Ukraine has increased the number of injuries and deaths since deployment; in year three there have been more casualties and deaths than in the two previous years using conventional weapons. In some individual battles the drones have been responsible for up to 80% of casualties and deaths. More than one million soldiers died by March 2025.

The Ukrainians started to deploy LAWs in their war with Russia, and these tools were responsible for early gains, which shows that a nimble tech-savvy smaller country can outmaneuver traditional armies and can usher in an age of killer robots. This is particularly important for a small country like Ukraine, which simply doesn't have enough fighters to lose. The article "Battlefield Drone and the Accelerating Autonomous Arms Race in Ukraine" describes it well,[13] and the Pentagon has expressed a desire to take deliveries of "one-way drones" that once equipped with a target go to hunt the target and kill the person or destroy the building/equipment. In the past there has been a specific rule for the use of drones by the US military, which leaves a human in the loop, that is, they make the decision to kill; the use of autonomous preprogrammed AI would be a major departure.

We tend to think of drones as airborne and large, but they can also be very small, like hummingbirds and vehicles used on land and sea.[14]

How has Ukraine been using them?

- They are largely off-the-shelf drones that are turned into LAWs in the theatre of war as needed.
- The mines, tanks, anti-tank defenses, trenches, and artillery, like mortars of traditional warfare, have given way to weaponized drones, often store bought, in attack or surveillance.
- Much of the war effort is now in minimizing drone attacks from either side by jamming signals and hacking.

- The standard arms are now covered with anti-drone netting.
- An example of such weapons—dragon drones—can spit out molten metal at 4,400° Fahrenheit over enemy lines before crashing.
- As reported by the *New York Times*, "Soldiers say it has become strangely personal, as buzzing robots hunt specific cars or even individual soldiers. It is, they say, a feeling of a thousand snipers in the sky." "You can hide from artillery," said Bohdan, a deputy commander with the Ukraine National Police Brigade. But drones, he said, "are a different kind of nightmare."[15]
- The sheer magnitude of the devices in use and development is mind-boggling. In 2024 the *New York Times* claimed that Russia made 4,000 drones a day and Ukrainian had made one million.[16] These are first-person-view drones, not autonomous weapons, but each combatant country plan to make three to four million of these drones in 2025. A first-person view drone can be guided to the target remotely by a human.
- Both countries have started using LAW drones.[17]

Last, and perhaps most important, all countries have noticed the way in which warfare is changing. Drones have been fitted with shotguns to fight other drones, small antiaircraft drones destroy surveillance drones flying higher in the sky, and large drones are expected to be developed to be launching stations for swarms of small drones, thus increasing the distance they can go in order to kill. At this point you might like to go back and watch the film *Slaughterbots*, which was made as a fictional story of what might come. Already, the Ukrainian Dovbush T10 can drop grenades on the battlefield or even carry small first-person view attack drones.[18]

The article "Battlefield Drone and the Accelerating Autonomous Arms Race in Ukraine" raises another potential issue with LAWs: they are often dual use technology and so can be weaponized even in their civilian context. A couple of examples:

- An autonomous drone that takes footage of you as you race through the woods on your bike is available as cheaply as $700. That drone could equally well be programmed to stalk and harass and frighten another person. Likewise, it could be armed and kill from a distance.

♦ Spot the Dog is an extremely sophisticated piece of robotics produced by Boston Dynamics. As we saw in Chapter 1, it can be used to help humans carry materials in difficult areas, for example, building sites. It can also be used as a guard dog. Indeed, in 2019 the Massachusetts Police Department became the first deployment in a civil setting of Spot as a guard dog. The Secret Service was seen using Spot at Mar-a-Lago in 2024.[19] In neither case was the robot armed but it could be and by someone with limited knowledge of AI or robotics. The robot is both manually guided and autonomous. Spot's price tag is upwards of $70,000 but a similar product can be purchased for about $3,000 from China via a popular online sales site. For law-abiding owners it might be the next stage in hypoallergenic pets, but for illegal use the possibilities are many.

In the United States one of the problems of military weaponry is that it has a direct secondary use in the civilian police force. As mentioned, Massachusetts has been using Spot since 2019, an almost immediate trickle down of an item made for the military market. How do you feel about having LAWs policing your town? Is a human police officer a necessity for you? As with military personnel, deploying robots in dangerous situations might save lives.

What Are the Worries About the Operation of LAWs?

The Future of Life Institute suggests the following five serious problems:

♦ **Accountability gap.** Autonomous weapons may kill more or different people than intended. Judges may therefore have difficulty assigning responsibility for war crimes.

♦ **Global instability.** Autonomous weapons are usually trained on classified data and may interact with enemy systems in unexpected ways. An unexpected movement by an enemy system can therefore cause "flash wars" (unintended escalation) without human override. Moreover, low production costs make them attractive to non-state armed groups as tools for genocides. They can also make it significantly easier to execute targeted killings of military or political leaders.

- **Ethics.** Key religious leaders, including the Pope, argue that machines should not be allowed to take life and death decisions. When polled, the vast majority of people reject this practice as unethical.
- **Vulnerability to cyberattack.** Autonomous weapons are uniquely susceptible to cyberattacks, creating new ways for hackers to infiltrate and manipulate military operations.
- **Unpredictability.** Autonomous weapons operate based on environmental stimuli, making their behavior uncontrollable and unpredictable.[20]

There are other potential problems from the use of LAWs, whether in military or civilian settings:

- AI doesn't recognize a person; it recognizes the target as collections of data inputs. Without constraints there is likely to be wider injuries to civilians.
- Two of the inputs used to identify the target are facial recognition and gait recognition. Facial recognition is known to be more accurate in identifying white faces than any other race. Notoriously, in 2015, Google misidentified black people as gorillas and was still having problems fixing it in 2018.[21] This is a simple data issue. There were more photos of white men in the algorithms training data than of any other person, including white women. Currently, 98% accuracy is claimed for facial recognition models. However, the methodology of coming up with such test scores has been the subject of debate.[22] Additionally, the testing has been done by National Institute of Standards and Technology, which is being cut by the Trump administration and could leave algorithms untested. In February the incumbent CEO of Clearview AI, a facial recognition company, resigned as new leadership joined the company to focus on opportunities presented by the Trump administration.[23]

- Lives have been upended by inaccurate facial recognition. This would be much worse if lives are ended as a result. And we have already seen facial recognition regularly used in warfare. The Israelis, during their current conflict in Gaza, have developed the ability to recognize part of a face. This means that civilian protesters would need to totally cover their face not to be identifiable through facial recognition.

- Gait recognition was spearheaded by China during the riots in Hong Kong.[24] Protesters baffled facial recognition by wearing face coverings. Each person has a different gait, which is easy for AI to learn from because it is excellent at pattern matching. Thus, it can spot images of a person walking in one setting and find that person walking in another setting with a 98% accuracy. It also doesn't suffer from the same racial problems as facial recognition and is therefore an excellent surveillance tool.

- In its "Antiqua et Nova" the Vatican expressed worries that LAWs could get out of the control of humans and be catastrophic for our future.[25]

- There are a slew of movies that show such scenarios where AI weapons get beyond human control. They include The Terminator series and *2001: A Space Odessey*. Many scientists also worry about this problem, and it is a considerable spur to the call for a ban. Additionally, many people worry that technology should be used to promote human well-being, which, they say, is not the case with LAWs.

- As with all modern AI systems humans are unable to interrogate them to see why the machine came to a particular decision.

- This will leave no room for the laws of war to operate when a killing, even a killing that was not authorized, occurs by a LAW. There will simply be no evidence of how it came to a decision. Likewise, if the LAW doesn't even know it's killing a person, who would be responsible for an accidental death or for using LAWs to commit atrocities? There will need to be new articles of the Geneva Convention in order to address these systems.

- Particularly, when LAWs are used in a civilian context there is a severe chance of algorithmic biases causing harm.

When data is lacking or biased, the machine will re-entrench the biases of our current societies.

♦ There is no doubt that LAWs will become regularly used by the militaries of many countries, and by their police forces, too. What is needed are thoughtful guardrails for their use so they cannot create untold harms.

Through LAWs AI will be involved in the military. The Pentagon said in March 2025 that they would be looking into having AI control nuclear weapons. It's worth repeating that the need for safety and security of AI could not be greater in this use. It was one of the ways in which the three scientists mentioned at the beginning of the chapter envisaged AI bringing about the end of humanity.

CHAPTER 6

The Justice

And then the justice,
   In fair round belly with good capon lin'd,
   With eyes severe and beard of formal cut,
   Full of wise saws and modern instances;
   And so he plays his part.
                    —William Shakespeare, *As You Like It*

Shakespeare uses "the Justice" as shorthand for the stage of life where a person is working. Therefore, in this chapter we will examine how AI affects our jobs and the hunt for a job, as well as other employment-related topics.

In January 2024 in an MIT report the International Monetary Fund stated that 40% of jobs globally would be affected by AI.[1]

In advanced economies, about 60 percent of jobs may be impacted by AI. Roughly half the exposed jobs may benefit from AI integration, enhancing productivity. For the other half, AI applications may execute key tasks currently performed by humans, which could lower labor demand, leading to lower wages and reduced hiring. In the most extreme cases, some of these jobs may disappear.

In emerging markets and low-income countries, by contrast, AI exposure is expected to be 40% and 26%, respectively. These findings suggest emerging markets and developing economies face fewer immediate disruptions from AI. At the same time, many of these countries don't have the infrastructure or skilled workforces to harness the benefits of AI, raising the risk that over time that the technology could worsen inequality among nations. Additionally, many countries in these two areas have very large populations of youth. If AI takes away jobs from young people eventually these countries could be dramatically and catastrophically affected.

In sub-Saharan Africa 70% of the population is under 30. This presents a huge benefit or a huge problem. If these young people can be educated in AI and then use it to create new jobs, develop new skills, and create the companies and jobs of the future, then that human capital could change the fortunes of countries that have been historically poor and reliant on selling minerals and other goods to the Global North[2] and subsistence farming. The benefits would not only help humans but also the environment because subsistence farming is a major cause of deforestation and the loss of habitat for nature in most countries of the Global South.[3]

The MIT report continues:

In most scenarios, AI will likely worsen overall inequality, a troubling trend that policymakers must proactively address to prevent the technology from further stoking social tensions. It is crucial for countries to establish comprehensive social safety nets and offer retraining programs for vulnerable workers. In doing so, we can make the AI transition more inclusive, protecting livelihoods and curbing inequality.[4]

Jobs of the Future and Losses to AI

Since Frey and Osbourne published their paper on jobs that might be lost to AI way back in 2013 it has been a topic of significant conversation and forecasting.[5] What we know is that some jobs will go away and others will appear. There are three issues worth discussing here:

- Jobs are not one thing. Each job is made up of a number of tasks. For a job to be completely lost all those tasks have to be capable of being done by AI. Of course there may be some jobs lost because some of those tasks can be automated and so the remaining tasks can be combined to be done by fewer people.

- How long is it between job losses and job gains? It took about one hundred years for the Industrial Revolution to bring benefits to everyone. During that time many suffered job losses and changes in how they worked. These revolutionary changes spurred the Luddite movement, a protest against the new automated machinery. These textile workers destroyed the new machines that they thought were taking their jobs. Indeed, the machines were taking their jobs and fewer people were needed to work in textiles until the automation of the jobs created such demand and opened new markets that more workers were needed to do new jobs.[6]

- Will AI unfairly affect jobs done by women? When the personal computer was developed executives were expected to type for themselves, and the majority of secretaries who took dictation and typed up the work for the executive almost totally disappeared. Secretaries were almost all women. Women have been very successful in clerical, customer-facing, and human interaction jobs. These are all highly likely to be automated by AI. This would have a catastrophic effect not only on the women whose jobs disappear but also the families supported by that work. In 2022, UNESCO first reported on AI and women's jobs. They updated the report in 2024.[7]

We know from pilots, for example, that repeated skills practice is required to handle situations when technical things go wrong. An US Air Force pilot does an average of 16 to 36 hours of simulation training each year. As we begin to adopt AI in the workplace we need to ensure that everyone gets regular training so that they are not relying, unquestionably, on the computer. For example, judges who are using sentencing algorithms should train often so that they retain their own judgment skills rather than relying on the algorithm for the answer. Of course, this raises the question: as our jobs move toward being handled by AI, how do we keep trained sufficiently to ensure it is working properly? If the act of legal judgment passes to an AI and the human judge ceases to exist who, beyond perhaps computer engineers, will try to ensure that the machine's judgments are correct? To whom do you appeal if an AI is wrong—another AI in a higher court? Indeed what will AI do to the court system which, in the United States, is an integral part of the checks and balances of the Constitution.

In June 2025, Dario Amodei, the CEO of Anthropic said, "AI could wipe out *half* of all entry-level white-collar jobs—and spike unemployment to 10–20% in the next one to five years." He went on to predict the "mass elimination of jobs across technology, finance, law, consulting and other white-collar professions, especially entry-level gigs." We saw how technology and coding was one of the "jobs of the future" but at Anthropic the AI itself does most of the coding, which is expected across the sector. Amodei is particularly worried that "lawmakers don't get it or don't believe it. CEOs are afraid to talk about it. Many workers won't realize the risks posed by the possible job apocalypse—until after it hits." He added that most employees "are unaware that this is about to happen. It sounds crazy, and people just don't believe it." And he foresees a world where "cancer is cured, the economy grows at 10% a year, the budget is balanced—and 20% of people don't have jobs."[8]

With AI doing a lot of its own coding, companies have been shedding paid tech jobs:

> The tech layoff wave is still kicking in 2025. Last year saw more than 150,000 job cuts across 549 companies, according to independent layoffs tracker Layoffs.fyi. So far this year, more than 22,000 workers have been the victim of reductions across the tech industry, with a staggering 16,084 cuts taking place in February alone.[9]

Steve Bannon, a former top aide to Trump and now a podcaster also told Axios that "AI will be a major issue in the 2028 presidential campaign." And President Obama has been speaking and posting about job loss from AI.[10]

Should There Be Human–Only Jobs?

Let's start with death. For centuries, humanity has experienced loved ones dying in wars and large-scale epidemics like the Spanish flu, which killed an estimated 50 million people worldwide.[11]

Drew Gilpin Faust wrote about how families dealt with the fact that their sons could not have "a Good Death" in the American Civil War because there was no one there to give them the rites of passage that had become formalized in the United States in the 19th century.[12] But gradually, in the Global North, as we fought wars in foreign lands and as more people died in the hospital, we have become increasingly separated from the moment of death. For many this heartache was exacerbated during COVID when family members had to sit outside hospitals in the parking lot while their loved ones died inside. Additionally, through news and films we may have become inured to death. But doctors—the ER doctor, the oncologist, the heart surgeon, and increasingly the obstetrician—have to deal with death all the time.

While modern medicine saves many more lives physicians still have to tell people that they will die. Could it be that AI will enable us to come closer to understanding, and living through, our own death? Would we be willing to have an AI tell us that we cannot be saved, or help us into death as assisted suicide becomes more common across the world? Even though AI might appear empathic in these situations we humans know that it is not and that it has no concept of death, being just a string of code.

Currently, it's likely that most humans would choose to be told that they are going to die by a human but in his *New York Times* article in October 2024 Jonathan Reisman wrote that "it doesn't actually matter if doctors feel compassion or empathy toward patients; it only matters if they act like it. In much the same way, it doesn't matter that AI has no idea what we, or it, are even talking about" (provided they appear to and act empathically). He went so far as to say in order to tell people they will die, doctors learn a script in medical school, and so a seemingly empathic AI could also use the script to give the news to a person.[13]

So as we move forward with AI in our lives how do we communicate what we want and ensure our wishes for care are known? Just as we carry living wills should we also carry instructions to emergency medical services about how we would like to have treatment from humans or AI, and will we even have a choice as hospitals and physicians use more and more AI in order to cut costs? What is the trade-off between savings and care, and how do we humans get a say?

Let's look at how AI affects different industries, starting with oncology.

ONCOLOGY

In 2023, I went for a routine mammogram. In the United States one in eight women will get breast cancer, and so it was for me. The routine mammogram revealed a tumor so small it could not have been found by human methods. As we know AI is excellent at pattern matching and I was pleased to have my mammogram read by both human and AI. When my oncologist discovered I knew about AI the oncologist said that she thought AI could be well used to help patients on their journey with cancer. I responded that I would rather that the oncologist helped me with my journey with cancer and that AI was used to free up the oncologist's time so that she could be with patients. However, there have been people who, on hearing this story, have said they would prefer help from an AI than a human because their doctor was terrible (either in ability or empathy). Another occasion when the AI might be very helpful is when the oncologist speaks a different language from the patient. The translation would give the patient the tools needed to ensure they understood and consented to their care and asked the right questions. Indeed in a recent report from Microsoft, they identified the jobs of translators as having the highest probability of disappearing to AI.[14]

It will always be a personal choice but the doctor and patient relationship may be one place where we want to keep the human.

RELIGION

We all know the hymn "Amazing Grace" but many of us don't know the back story. The author was a slave trader who after a very stormy Atlantic crossing was washed up alive on a beach. His hymn celebrates his survival and thanks God for it. Following this incident he not only became a cleric but also invigorated William Wilberforce, whose work in the Houses of Parliament eventually led to the abolition of the slave trade by British citizens or British-flagged ships in 1807 and its colonies in 1833.

Does inspiration to do the work of your God come from the human or can it be augmented by the machine? The immediate response might be the human because machines don't have a soul, but a number of religious leaders have been experimenting with AI to help them write their sermons and appeal to a younger audience.[15]

The Vatican has taken an interest in the development of AI and how humans fit into the future for some time. It has convened groups, including the representatives of AI companies, governments, nonprofits, faith leaders, and academics. Most recently, in January 2025 the Vatican released "Antiqua et Nova (Old and New): Note on the Relationship Between Artificial Intelligence and Human Intelligence." It recommends that AI should always be a tool that assists human intelligence but does not replace it, and emphasizes that there are unique aspects of being human.[16] The newly elected Pope Leo XIV has stated that AI is at the top of his list of issues that the Catholic Church should address.[17] In June 2025 he made the threat of AI to humanity a signature issue of his papacy.[18]

Other faith leaders have also come together to discuss how AI can be balanced to promote humanity. In 2024, leaders from the Episcopal Church, the Presbyterian Church, the Evangelical Lutheran Church in America, the Catholic Church, The Orthodox Church, the United Methodist Church, and the Evangelical Church came together to discuss the topic under the auspices of a nonprofit "AIandFaith".[19] AI is widely used and supported across the Muslim world but with an emphasis from religious leaders on the good of humanity.[20] The conversations in Orthodox Jewish households are complex, especially where use of computers is banned on the Sabbath.[21] But we have seen that AI has been widely used by the government of Israel in warfare and some of the most enterprising startups are located in Israel. In the Hindu faith there is concern about "the erosion of human autonomy and the loss of control over decision-making processes,"[22] and in Buddhism there is a similar concern about how humans can co-exist with AI and not have their spirituality harmed by it.[23]

LAW

As mentioned, AI hallucinations are becoming a common problem encountered by lawyers and the court system. Increasingly lawyers and litigants in person are asking AI to draft legal documents. There is a lot of anecdotal evidence of cases cited in legal documents being made up by AI or where the name of the case is correct but facts have made up and have nothing to do with the original case.[24] This means that humans have to check and double-check everything they send out. In the United States, courts can dismiss a case in these circumstances or fine or hold criminally liable the lawyer responsible for submitting inaccurate AI-produced documents. Obviously this is a huge risk for not only the lawyers but also the litigants, who may find their case dismissed and may have to sue the lawyer for professional negligence. In the United Kingdom, the courts have advised that not only will the actual lawyer filing the document be held in contempt but also the head of that lawyer's company.[25]

Also in the United Kingdom, their senior judge has questioned the use of AI in researching and bringing cases to Court. He suggests that lawyers could be negligent by using AI but also in not using AI if the case could be helped by detailed analysis only an AI tool could do.[26]

In future law firms might want to ensure that their clients understand the benefits and risks of using AI before embarking on litigation. It is necessary for clients to insist that someone who is qualified checks any documentation filed with a court.

Because this problem will get worse. Any data, correct or incorrect, created by AI becomes accessible to use in the future. This applies to all knowledge searched by an AI but is a good example to illustrate the problem of hallucination. For example, if the AI invents, or alters, 17 cases for a plaintiff and then another 17 for the defendant there are 34 invented or altered cases for it to choose from the next time it is asked for cases on this topic. It could just use one of those 34 or invent new fake cases based on them. Soon it is very hard for lawyers and judges to know what is true and what is not without employing sophisticated tools, possibly more AI. This leads to a contention by AI developers that we can constrain the behaviors of one AI by creating a "guardian AI" to watch over them and correct their behaviors. It's hard not to believe that this is simply creating another AI lawyer, which will have to be supervised, as there is no guarantee it will hallucinate less than the one which it is overseeing.[27]

It is often said that AI would make a better judge than a human. Those who say this see the act of judging as black and white, whereas, as a judge myself, I can say categorically that there are many shade of gray in all legal cases. Often one side doesn't have all the right, and it is up to the judge to use their experience and humanity to make the right decision. It could be said this follows a tradition set by Soloman, the king in the Old Testament.[28] The course of justice runs differently in various countries and their legal systems. For example although the United States and United Kingdom share a "common law" system the former elects many judges while the latter uses a rigorous interview and examination process. In Europe and other countries using the "civil law" system, in which students choose to study either to be a lawyer or a judge. Whether AI can ever reach a point where it truly understands how humans make judgments is unknown. Our common sense of morality is something OpenAI has funded Duke University researchers ($1 million over three years) to discover.[29] It's interesting to think that OpenAI has spent billions on creating AI but only funded $1 million on research into how humans make complex moral judgments in high-stakes areas like medicine and law. Given the Global North, US-centric biases of the data ingested by AI, it remains to be seen if this can be done.

Further issues will develop if AI is used as a judge in the future. For example, who and how will it be possible to appeal an AI judge. The human judge writes a judgment, which can be appealed. While the appeals court reads the judgment, arguably they could ask the human judge to explain their reasoning. That's not possible with AI systems where systems have a lack of explainability and transparency. Additionally we should weigh into our decision-making that AI systems, as we have seen, cannot rise above hallucinations. Should we use systems that might make up significant issues in a legal case? So the question here is not whether we use AI systems now but should we use them when we know they will never not hallucinate? Also, who will hear an appeal from an AI judge? An AI, a group of Ais, or humans? Should we preserve a role for human beings in our justice system? Do we want AI to make determinations about how humans conduct their business and, especially, in the criminal justice system?

What is at stake is, for common law jurisdictions, a total breakdown on the legal system. Judges in common law jurisdictions make judgments based on cases that are similar to the one they are deciding that act as a precedent for the way the case should be decided. Ultimately the highest court in the land can change the president (as we saw in *Rowe v. Wade*) in the United States. If the use of AI prevents judges being able to trust the cases that are being put forward as precedent then the system will grind to a halt, or worse still, decisions will be made on completely false precedents. For modern society to work the rule of law must be unquestionable.

HUMAN RESOURCES

As we have seen job losses in tech are continuing but also in other sectors implementing AI. For example, banker Morgan Stanley cut 2,000 jobs specifically due to AI[30] and the banking sector is likely to see 200,000 jobs lost over the next few years.[31] In 2024, CITI's researchers said that 67% of jobs in banking have the potential to be automated.[32] The immediate effects of highly paid workers losing their job ripple across others whom they employ, for example, maids, garden services, drivers, and more. It also affects the housing market if they have to sell their home, and private schooling if their children have to be withdrawn, which in turn affects the teachers employed there. When higher paid workers lose their jobs the effect ripples through the market for those who are paid less because their livelihoods depend on the purchasing power of the highly paid employee. Additionally, it is important for companies to ensure diversity is retained in the company during these lay-offs.

Getting a job, let alone getting the job of your dreams, is one of the most important things we do as adults. Many young people feel that they are the most rejected generation because they have to submit hundreds of job applications without often getting even a response from the potential employer.[33] For many of those individuals the receiver of their applications is not a human being but an AI. Even before the advent of generative AI, human resources (HR) was one of the early users of AI. In large companies they have to either rely on AI or take on many extra human beings to do the work. For example Goldman Sachs has 2,700 internship positions and receives roughly 315,000 applicants, which means that about 312,300 get rejected.[34] Many companies outsource their AI hiring tasks to companies like Workday but others are buying from early-stage startups or creating them in-house. Additionally "people analytics" in HR departments are growing. Because they are dealing with the data of employees, it is of vital importance that they adopt good privacy and AI governance practices.

There are three major reasons why AI is used in HR:

◆ Efficiency
◆ Personalization of experience (although it is often hard to see how using a machine would personalize a human's interaction with HR)
◆ Task automation

The sort of tasks that you might find HR doing with help from AI or simply handing off to AI to do include the following:

◆ Recruitment (AI can help in writing job descriptions, screening applications, communicating with applicants, scheduling interviews, conducting initial interviews, selecting candidates, in other words, nearly the whole process. Therefore, it's important that applicants understand and compensate for the use of AI. As we know, AI is a pattern-matching tool and so cover letters/CVs/résumés should ensure that they have keywords and phrases in them that replicate those in the job description. It might even be wise to get generative AI to write these documents, provided that it is not banned in the application.[35] I recently reviewed an application that had clearly been written by a chatbot, which was in complete disregard for the rules for the application. Needless to say the candidate was not selected. Recruitment is very expensive for companies, especially when they are recruiting many people daily. AI serves as a tool to help them recruit more efficiently and more speedily.) As we are in the early stages of this selection by AI we don't have good data as to whether AI makes the right selections

- ◆ Tools for checking efficiency, performance monitoring, personalizing development training, identifying candidates who need training, enabling managers to give AI-created or suggested feedback, setting goals for employees
- ◆ Administration, such as benefits, record keeping, payroll processing, and more
- ◆ Onboarding

Among the benefits touted for the use of AI in HR is that it can do the following:

- ◆ Improve the candidates experience
- ◆ Improve the employees experience
- ◆ Assist in training and identifying potential employees for training and leadership

Recently some companies have moved away from the title Human Resources in favor of the name People and Culture or People Operations or Employee Experience. The idea is to help employees know that they come first as human beings in the company. Unfortunately, excessive use of AI in this area can undermine this ethos. Although it will help the manager enormously to have AI write employee appraisals and training suggestions, employees may wish that their manager had prioritized their time to show that they valued the employee enough to think about these things themselves. Likewise, candidates have felt as if they were talking to Siri during interviews with AI and worry that they had not been chosen because their application didn't meet the needs of the AI, even if they were more than qualified for the job.[36]

Unfortunately, those who are turned down for a job by AI systems cannot be given detailed feedback. That is because of a lack of transparency and explainability about how the machine came to its decision. This is one of the most high-risk uses of AI. It is defined as such by the EU AI Act, and several states in the United States have laws that attempt to constrain the use of AI in HR, and allows a rejected candidate to ask for reasons why the AI did not choose them. The first place things go wrong in HR is the choice of training data for the model. In one famous case a bank trained its algorithm to search for people with similar profiles to those who had been successful in the bank. This ended up with the only candidates being chosen being white men from Ivy League Schools because that was the historic profile of people who had succeeded at the bank. So, like all AI tools, an HR system needs to be trained to ensure the best candidates are chosen regardless of gender, race, and more, and there needs to be an appeals process for the candidate who may have been harmed by the AI. At any time when you are seeking a job, find out if AI will be used. You can then interrogate, and possibly sue, the company if you don't get an explanation for why the AI didn't choose you.

Likewise when a company is looking to promote employees, HR professionals will need to invent new ways of finding leaders. If everyone in the company has the superpower of using generative AI how should your company pick the exceptional human from the pack to train to lead? Is it the person who can best use AI or is it the person who thinks creatively and strategically and uses AI. What are the features by which an HR professional or a manager can spot exceptional talent?

AI Surveillance in Jobs

Your employer can use AI to watch every move and every click that you make. This puts every worker under constant surveillance. As with all things AI, there is a positive and negative side. In dangerous jobs AI can help to keep workers safe by spotting when the worker does something wrong or is going to do so. In fact, robots enabled with AI are being used to protect human workers. Examples come from many work sectors, but bomb disposal is of huge danger to humans and now AI can take that on.[37]

On the opposite side, PWC developed a tool that uses facial recognition to check when employees are in front of their computers.[38]

Financial service companies might need to monitor their workers, but used generally by employers it could feel very intrusive and privacy invading for employees to be constantly monitored by "big brother." I was recently asked for help by an employee whose employer used keystroke tracking to check whether she was working. She was so worried about it that when she went for an emergency dental appointment she took her computer so her keystrokes for that day would be the correct number required of her.

Human Trafficking and AI

Human slavery is the scourge of many societies, with young and old forced into labor without pay around the world, even in countries in the Global North. The United States is believed to have some 300,000 individuals working in the sex industry as slaves. In this short overview I spoke to Ann Ewasechko, who works with Hewlett Packard Enterprises[39] on their initiative to stop human trafficking in supply chains. She has this to tell us about human trafficking and AI for good:

> As we have seen AI is nothing without data. A revolutionary shift in technological solutions is taking place where it is now possible to share AI training, learnings and insights without sharing sensitive or individual private data.
>
> The impact of such protected data sharing shone through during the COVID-19 pandemic. When pharmaceutical companies pooled their data, the scientific community was able to create effective vaccines, test them for efficacy and safety, and distribute them globally in record time. This is an example of AI working at scale and technology bringing a global good.
>
> Similar impacts are seen with research into Alzheimer's, which requires collecting exponential amounts of data on patients and using AI and supercomputing to analyze it. Information from genomics, brain imaging, and clinical studies from multiple sources must be accessed and analyzed securely to protect patient privacy.

Sharing data across researchers is crucial, but what if the true impact comes from leveraging data from multiple sources worldwide—governments, businesses, suppliers, academia, and NGOs, where trust and understanding have not fully taken form? Take the issue of forced labor, which fits squarely in the intersect of business, government, and humanity.

Forced labor is a grave human rights violation where individuals are coerced into work under threat, deception, or abuse. In 2021, an estimated 27.6 million people were subjected to forced labor globally on any given day.

The impact of forced labor is profound and multifaceted. Victims endure physical and psychological trauma, including violence, abuse, and deprivation. Foreign migrant workers are also particularly vulnerable to huge financial burdens from onerous recruitment fees. Economically, forced labor generates an estimated $236 billion in illegal profits annually, fueling criminal networks and perpetuating cycles of exploitation. Socially, it undermines the dignity and freedom of individuals, often entrenching poverty and inequality. Despite international efforts, the prevalence of forced labor remains alarmingly high, necessitating urgent and sustained action to eradicate this grave injustice.

Many companies do not believe that their profits are based on forced labor. It is not because they are intentionally exploiting labor; it is because they do not know. Telling businesses to dig deeper into their supply chains, governments to enforce more, or nongovernmental organizations (NGOs) to save more people is not ultimately going to win the day.

HOW DOES AI AFFECT FORCED LABOR?

AI tools are increasing being used by enhancing detection, prevention, and support mechanisms. Facial recognition and behavioral analytics identify and track victims of human trafficking, aiding law enforcement. AI-driven analysis of satellite imagery assists in pinpointing regions vulnerable to forced labor—such as brick kilns and shipping patterns—enabling targeted interventions by NGOs and authorities.

The potential for AI is so much more. There is considerable data that organizations collect to identify, monitor, and mitigate risks within global supply chains. Corporations collect data on employment practices, recruitment fees, and employment contracts, and conduct audits and assessments to verify compliance with standards.

NGOs conduct extensive surveys, interviews with workers and community leaders, and on-site intelligence gathering to pinpoint vulnerable individuals and estimate the prevalence of modern slavery. Often through worker voice tolls, they assess worker conditions, document audit results, and track corrective actions. Governments collect data on compliance rates; contents of statements, goods, and services; source countries; and cases that move through their support processes. The United Nations collects data on forced labor through surveys, reports, and collaboration with governments and organizations, such as Walk Free.[40] They gather case data, conduct research, and partner with member states to improve data collection

Unfortunately, most of the data is siloed and remains so without creating the impact it could—it is trapped in every organization. Imagine how much more effective organizations working to combat modern slavery could be if it were possible to unlock and democratize data trapped in every organization, making connections from the village to the boardroom, involving everyone in the chain.

When this is done and the power of AI is brought to bear, there will be no place for exploiters to hide. Modern slavery hides in plain sight because it is able to hide. These technologies can help achieve this, and there is excitement about the possibility that technology can play a role in finally winning the war against modern slavery.

In some ways AI can be thought of as a hammer; its use in human trafficking is a great example. As described by Ann this use of AI would be akin to the way a hammer has been designed: to create things by inserting nails. On the other side of this equation, the hammer can be used to maim and kill. So too with AI. Traffickers use AI in many ways to entrap victims.[41] Some even think that using AI to fight trafficking is dangerous.[42] It is well known that predators use social media to meet, groom, and enslave sex slaves.[43]

To discover more about human trafficking, aka modern slavery, you can read my book *Human Trafficking and Human Rights*.[44]

Is Creativity Solely a Human Skill?

Creativity was thought at one time to be the last bastion left to humans in the age of AI. However, we have gradually seen that AI can create all types of design, music, paintings, video, pictures, prose, and poetry. For many this is a cause for concern because we have always believed that creativity comes from within us; it was part of what made us human. But now those without artist prowess can create art. Additionally, those who have a disability are given access to creating something in a style they love. AI brings us new ways of showing our creativity just as technology has helped in the past. Weavers would not be able to create wonderful materials without the loom, which first made its appearance in about the 5th century BCE. In 1785, at the start of the Industrial Revolution, Edmund Cartwright created the first mechanized loom by using steam power, enabling many more people to make cloth. At the same time, this advancement could be thought of as taking away human creativity, turning us instead into the makers of predetermined cloth and the "watchers" of machines. Is it possible that AI will bring about the same change in, for example, art. We humans give the machine a prompt and then watch while it creates the image. The human can then tweak the image endlessly but the human no longer is the creator in the way they were with pastels or oil. Likewise, instead of painfully writing this book from all my ideas, I could have watched a chatbot create the book, made some changes, and then put my name of the cover. But is that my creation? Just as whether we respect the rights of human beings, when we think about the need for responsible AI development, so too should we value the rights of humans to sustain their creativity. Does the need to go somewhere, even into the yard, to take a photo change our relationship with the object and open new horizons for us as humans. Can we truly relate to nature if we don't interact with it by painting, writing, or photographing it. Composers have often celebrated the countryside through music; composer Gustav Holst also celebrated the planets. Can we as humans get as much out of creativity if we don't interact with the subject?

Gabby Salazar is a National Geographic explorer, a conservation photographer, and an environmental social scientist.[45] She picked up a camera as a teen and was entranced by the beauty of nature all around her. In 2004 she was named BBC Young Wildlife Photographer of the Year and since then has dedicated her life to photography, particularly conservation photography.

Asked to think about her creative art in the age of AI Salazar said that during the time she has been a photographer she has seen many new technologies "from digital photography to drones." Rather than seeing them as a threat to her career she has "embraced these technologies as new tools that can enhance my creativity." She is thinking about AI in the same way: as a suite of tools that can help her do her job better and extend the creative process. She uses "chatbots for brainstorming and I use filters on my iPhone, such as Portrait Mode, to blur backgrounds."

However, there are areas where she deliberately lets a photo speak for itself, untouched by AI. As a photojournalist her job is to "document the real world through images and my work needs to meet high journalistic standards."

She thinks it is a different thought process for artists. Their goal is the creative process and

> AI is being harnessed in new ways for creative expression and is leading to novel and interesting types of art. I think that there are exciting possibilities as people learn to more effectively utilize AI to realize their artistic visions. When I teach, I encourage my students to approach AI as another tool in their creative toolbox.

Nevertheless, Gabby's view is that there will always be a place for art that is created by a human. She points out that painting didn't die with the invention in 1839 of the daguerreotype. The daguerreotype was the first way of taking photographs and widely used in the 1840s and 1850s. It used a photographic process in which iodine-sensitized silvered plates and mercury vapor create the photograph. She goes further to explain that she is just one of many artists for whom the creative process brings pleasure and satisfaction:

> I simply love being outside using my camera to photograph the natural world and I feel happy when I create something new and beautiful. The images I create are also tied to my unique experiences and memories. For me, generating a similar image using a machine would not provide the same satisfaction or depth of feeling, nor would it give the same result.

One of the problems in using generative AI models is the quality of data available to the AI. For example, if there is no or little data about a spot in Borneo then the AI cannot create a quality image. As always, with AI the quality of the product depends on the data it can access. Therefore, Gabby says, an AI model is "not able to directly reflect reality in the same way that I can with my camera."

Does the creative process need time? Gabby thinks so:

> Philosophically, one of the aspects of generative AI that I find most antithetical to the creative process is its speed. Some programs can generate an image, painting, song, or poem faster than I can write this sentence. If my main goal is productivity, these programs are invaluable tools. But, if my goal is creative expression—to produce something of beauty or with a deeper meaning that is unique to my experience of the world—I need time. For me, creativity has always been a slow and deliberate process, involving reflection and discovery. When I think of slow creativity, I often think of the Austrian composer Gustav Mahler, who wrote some of my favorite pieces of music. To focus on his musical compositions, he retreated to "composing huts" in natural settings, first in Austria and later in Italy. There, he could shut out the world and find the silence he needed to develop his creative thoughts. He also found inspiration in the natural settings around the huts, which is reflected in many of his

symphonies and songs. Even if AI can produce an imitation of Mahler's work, for me it will lack the depth, meaning, and emotion that is interwoven in his original pieces.

She added that she is open to being surprised by new work.

In 2023, will-i-am spoke in a fireside chat in Davos about AI and music. He celebrated the fact that technological development had helped him have a career in music and he hoped that AI would enable other people, like him, to be inventive in the way that music is produced. Of course, it needs to be responsibly used and not merely copies of the work of others. He added that it's always interesting to hear how new works can be created with new tools. They change and empower those who might not have been able to be involved in the creative process before.

Another use of AI for films, which you could think of in terms of advanced computer-generated imagery, is repeating a retired or dead actor's image in new films in which they had not starred and with which they had no connection. The State of California passed laws to prevent this being done without express permission but most US states and other nations do not have such legislation. In other words we could still see Tom Cruise in Mission Impossible films in 2100. As an actor who appears to relish doing his own stunts and the human process of movie making, that may make him very uncomfortable indeed. Likewise, if we keep our movie superstars alive it limits the opportunities for young talent. Companies are now springing up that enable you to keep the rights to your image and digital footprint post death, or you could put a clause into your will.

There are many other creative disciplines for humans. Will an AI-enabled robot be able to challenge a Michelin-starred chef? We have yet to see but you might be using a Chatbot to suggest recipes for using up things from your refrigerator in a new recipe. If that saves your throwing out food then it might be a great thing for your purse and the planet, but always weigh up the environmental cost of asking a question of a chatbot with the benefit you are getting from the answer (see "Impact of AI on a Child's Future Environment" in Chapter 2).

Although there are few AI-enabled robots using oils and canvas with which to create art, human artists are exploring how to use AI to make a new form of original art. In 2023, Refik Anadol astounded visitors to the World Economic Forum's annual event in Davos with a huge AI installation. He was using AI to interpret and reinterpret his own photographs, taken of a coral reef.[46]

In 2025 the World Economic Forum chose Sougwen Chung's art installation SPECTRAL as a convergence of art, technology, and human creativity. She uses machines as co-creators of her art by using custom robotic units intricately connected to her brainwaves through an AI-driven program she developed. This transforms her biofeedback into a dynamic canvas, using an electroencephalogram headset to measure and analyze her brain activity. The machines create alternative marks, positioning Chung as a conduit for artistic expression. Chung believes that the way she works shows that drawing need not be a purely human creation but instead be a collaboration between the human brain and the AI. This could help us humans see new ways of working collaboratively with machines while keeping our own unique human identities. Chung, like me, was a 2024 *TIME Magazine* impact awardee for her work in AI.

DOES AI WANT TO BE CREATIVE?

AI doesn't actually "want" anything. It's just a set of codes that combine to do remarkable things. Should AI become superintelligence it is possible it would wish to have a creative side, perhaps with the ability to self-actuate in a way we humans consider equivalent to the art of Leonardo Da Vinci or maybe in a way in which other superintelligent AI systems would appreciate. But at the moment AI art is about humans prompting it to create art from their own experience.

In March 2025 OpenAI showed off its chatbots' ability to do creative writing, a 1,000-word composition called "AI and Grief."[47] As we have seen, the only way in which an AI can write anything is to draw on the collective knowledge of everything on the internet. Therefore, its work cannot be drawn from people whose data is not on the internet. Thus, the paper is limited to "grief" in parts of the world where the internet reaches. But, perhaps most important, it can create something only from past data, not create something new from a writer's personal human experience. As humans we write about and ask questions that others haven't thought of or didn't ask. This is what makes human art, music, scientific discovery, and music unique. AI simply cannot do that;

it can merely react to a prompt. So can we include that life experience in a prompt or should we just use AI as a great partner for brainstorming.

Likewise, when we humans react to art, are we also reacting to the humanity of the artist? Creation of anything is hard work for humans: the output reflects lives, our hopes and fears, and our experiences. It's hard to see how a machine, which has none of those experiences and draws on data mostly supplied by white men, can engage humans as intimately as the human artist does.

In my view, what we are seeing at the moment in AI is akin to 12-year-old children. They think they know a lot, and so they are very certain of their skills in many things, but actually they are on the cusp of learning the things that give us older humans the life experience that we recognize in the work of others.

MUSIC: VINYL OR DIGITAL?

For many years music enthusiasts wanted the most perfect experience they could from music systems. That perfect experience would be akin to the live performance. Now we have AI to create this perfect experience, it is interesting that there has been a rise in the purchase of vinyl records and purchase of hi-fi systems.

This discussion of whether creativity is unique to human beings goes to the very heart of our discussions about not only what AI *can* do but also what it *should* do. If AI becomes able to do all the things we humans can do, then how do we define a role for ourselves?

One way in which humans have been using AI creatively has been in the creation of deepfakes. These are usually made to the detriment of someone or society. So, sadly, we are already using the immense power of AI in a very unfortunate way!

The Deepfake Problem

A deepfake is made by using AI to show a human being doing something, saying something, or behaving in a way that they did not and probably would not ever do. Examples range from deepfake porn with Taylor Swift[48] to Pope Francis wearing a particular type of coat[49] to the viral video that may have changed the result of the elections in Argentina in May 2025.[50]

While we have seen deepfakes in the school setting, of celebrities, and in elections, businesses are now having to be concerned about them being used as, essentially, cyberattacks. In 2024, a bank worker in Hong Kong was asked to wire $25 million out of the company. The required procedure was for an online meeting to be convened with her colleagues whom she knew but were at remote locations. She convened the meeting but everyone in the meeting was a deepfake. She duly wired the money because the fakes were so excellent. Since this time deepfakes have gotten even better. For example, it used to be that they could only look at the screen so you could test by asking people to shake their heads. That's not a foolproof method anymore.[51]

To combat deepfakes the industry has started an initiative called C2PA. Led by Adobe it unifies the efforts of the Content Authenticity Initiative, which focuses on systems to provide context and a history for digital media, and Project Origin, a Microsoft- and BBC-led initiative that tackles disinformation in the digital news ecosystem. Although watermarking of images is not foolproof, it is currently one of the best ways to prevent deepfakes. The person creating the image introduces a watermark into the history of the image, which then stays with it every time the image is changed. This enables users to track back through the history of the image and see how it has been manipulated.[52]

Other ways of testing for deepfakes include the following:

- **Facial anti-spoofing.** Here, the human uses a specialist facial recognition product to look for inconsistent facial features or movements.[53]
- **Liveness cues detection.** Again, the human uses a specialist AI system that tracks signs, such as blinking patterns or subtle movements, that enable it to decide whether the person is human.[54]
- **Artifact and anomaly analysis.** The human uses AI to find the inconsistencies or artifacts in that media (data) that can show tampering or AI manipulation.
- **Deep learning architectures.** Users can buy a trained AI system, which can analyze media (data) to recognize patterns or characteristics that are likely to be deepfakes.[55]

Of course, this is really a case of AI development chasing after AI development. There may come a time when deepfakes are so good that we are unable to use these detection measures. In the meantime, there are a number of companies that provide tools to help business owners detect deepfakes, although threats change frequently not having such a protection is like the early days of cybersecurity, which led to many disasters. Also encourage employees to ask questions of others in video conferences to ensure they are humans and learn how to review photos to ensure their veracity. It may feel odd for employees to be asking each other questions that only the human knows the answer to but it may save a costly hack happening against your company.

Another area where we are seeing deepfakes used to foil humans is in legal cases. Litigants are creating deepfake doctors' reports and other expert evidence, which means that doctors and businesses should be taking steps to ensure that their documents cannot be deepfaked. In one case I heard that a doctor's office had put a mark on their documents to ensure their authenticity. The litigant merely created the authenticity mark in the deepfake, too.

In California, as of June 2025, there is legal protection for artists not to have their voice or likeness used without their consent and other legislation to combat deepfakes. California initiated these protections in 2020.[56]

However, for individuals it's a bit of a wild west in many countries (see Chapter 4 for more).

Up until now we have largely been looking at the way in which AI is affecting our society and our families. In Chapter 7 we move on to think about its impact on business and government, including our employment.

CHAPTER 7

The Pantaloon

The sixth age shifts
>   *Into the lean and slipper'd pantaloon,*
>   *With spectacles on nose and pouch on side;*
>   *His youthful hose, well sav'd, a world too wide*
>   *For his shrunk shank; and his big manly voice,*
>   *Turning again toward childish treble, pipes*
>   *And whistles in his sound.*
>   —William Shakespeare, *As You Like It*

As we grow in our career, our needs and position in the world change. Shakespeare's Pantaloon is defined as a silly old man, but in this chapter I use it as a metaphor for that stage of our life when we are more mature in our careers and lives.

So here is a good place to look at the benefits and problems of AI in our medical care, policing, business, and government for those using and not using AI and how AI might change the way in which the workforce develops in the future.

The Emergency Medical Hologram

In Star Trek: Voyager the doctor on the spaceship is a hologram generated from a computer that uses powerful AI tools to enable the hologram to act as a human. In the episode "Tattoo" the hologram specifically explains, "I don't have a life, I have a program."[1] As the series develops we watch the AI program develop human attributes. However, does this change make it human or is it still a program? The writers of Star Trek: Discovery imagined that by the time we reached 2024 the spaceship's computer would be accepted as sentient.

Currently, AI doctors are not available but AI-enabled devices are helping human doctors in all sorts of situations. Here are just a few:

- AI can monitor pacemakers, and more work in this area is to come.[2] However, these devices need to be hyper secure as they are essentially Internet of Things (IoT) devices. As such they can be hacked for the purposes of ransoms or even murder.
- For many years we have been told that radiologists won't have jobs in the future because of AI. It is certainly true that AI is excellent at reading X-rays (it's a pattern-matching tool after all); however, radiologists are still employed and will be until the public accepts, for example, their cancer diagnosis from an AI system.
- There are thousands of small AI companies looking for ways of improving health by using AI but one of the problems with providing health care using AI is the personal nature of the data and, rightly, the levels of security to protect that data. Also, as we have seen, problems with the quality of data and its applicability to the specific patient could hinder the positive use of AI.

So how can we find sufficient data to be enable AI to make excellent medical recommendations like the imaginary medical hologram?

- **Scholarly articles.** These articles are often free or of nominal cost. The scientists at DeepMind were able to create AlphaFold, which is an AI that can predict 3D structures of proteins with a high degree of accuracy by reviewing huge numbers of articles that would take a human more than their lifetime to read. This technology enables us to understand how the human body interacts with diseases.[3]
- **Anonymized data.** This data helps researchers to understand illnesses. Often people will donate their medical data for research purposes if they can be sure that it cannot be traced back to them. For example, if every woman with breast cancer

donated her data to be used by AI, better detection, individualized therapies, or even new cures might emerge.[4]

♦ **Data made available by nationalized health care services.** In 2023, the National Health Service (NHS) in the United Kingdom allowed Palantir, the company started by Peter Thiel, access to the medical records held by the NHS.[5] The NHS holds data on every person (most UK citizens) treated by it since records began. It is a treasure trove for any company wanting to create any cure, device, or operation with AI. However, it raises huge privacy concerns for patients. For example, as Palantir is an US-based company it is subject to US laws, which could lead to this data passing on to the US government.

♦ **Data vaults.** These vaults are controlled by nations who allow data to be deposited into them and then taken out, after anonymization, to be used by researchers. This anonymization protects the privacy of the patient while allowing researchers access to a large volume and variety of data.[6]

♦ **Digital twins.** By using sensors and other readings AI can create a digital twin of you and your illness. It can then experiment with different therapies to find the best one for you without subjecting you to any drugs or procedures.[7]

♦ **Data collaboratives.** As described by David Bray, one of the founders of the concept, a data collaborative functions as a trust-based framework where multiple stakeholders collaborate to share, manage, and derive value from their combined data assets while maintaining individual autonomy and privacy rights.[8] This model emphasizes democratic decision-making and equitable distribution of benefits derived from the collective data pool. The touted benefits are as follows:
  • Builds community resilience through shared data resources and expertise
  • Reduces individual costs of data management and analysis
  • Empowers members with greater control over their data's use and monetization

There are currently many uses of AI in health care and these will vastly increase in number once more data is available. Countries across the world are thinking about how AI can be used, especially where there are few human doctors per patient. One such country is Rwanda, where there is one human doctor for every 10,000 patients, well beyond the recommendations of the World Health Organization of one doctor for every 7,000 patients.[9] As a result the government of Rwanda has employed an AI provider to triage calls to doctors, which is aimed at ensuring patients in the most need get the quickest help. It was implemented with much fanfare in 2020, but it is hard to find out how it worked and whether it's still in use.[10]

There are many providers of AI mental health counselors.[11] If you are using this type of therapist, all the same considerations apply that we talked about in Chapter 4. Always remember that AI is a prediction machine, not a person. There needs to be more research into how successful AI counselors are because they are so new we have not been able to test them to ensure that they provide the same level of care as humans. There is anecdotal evidence that some veterans prefer speaking to the bot than a human.[12]

Doctors are required to keep copious notes of interactions with patients. Many companies now provide recordings and AI transcription of patient notes. If you would rather the doctor takes notes instead, ask your doctor's office if they are using an AI transcription device and then decline its use. The benefit of these tools is that they let the doctor have more human interaction with the patient and more observation time (rather than typing all the time as you describe your symptoms). However, we know AI is not yet perfect. Doctors should be checking and signing off on AI-transcribed notes. As a patient you may also want to double-check the notes correctly reflect your interaction.

The use of AI in health care in the United States is regulated by states (at the moment), the Federal Drug Administration (FDA), and the Department of Health and Human Services. In countries with socialized medicine the government of the country sets up an independent regulator, but hospitals and practices using AI make treating decisions.

On January 13, 2025, the Attorney General of California released legal advisories about civil rights, date privacy, and consumer protection concerning the use of AI, including in the health care context, noting specifically that businesses have obligations under these acts:

♦ California Consumer Privacy Act 2018
♦ The Student Online Personal Information Protection Act

◆ The California Invasion of Privacy Act
◆ The Confidentiality of Medical Information Act

The health care AI advisory is particularly important because if AI is wrongly used in medicine it can kill patients. The health care AI advisory provides guidance specific to healthcare providers, insurers, vendors, investors, and other healthcare entities about their obligations with respect to their use of AI under California law, including:

Health consumer protection laws (e.g., prohibition on unlawful, unfair or fraudulent business acts or practices; professional licensing standards and other prohibitions relating to the practice of medicine by non-human entities; requirements relating to management of health insurance); anti-discrimination laws (e.g., requirements relating to protected classifications); and patient privacy and autonomy laws (e.g., use and disclosure of patient data, confidentiality of patient data, patient consent, patient rights).[13]

Likewise, the FDA is the regulator for AI-enabled medical devices. In 2023, the agency issued a comprehensive list of approved devices, which added to the 2022 list. There were 637 devices[14] covering the following areas:

◆ Radiology: 531 devices
◆ Cardiovascular: 71 devices
◆ Neurology: 20 devices
◆ Hematology: 15 devices

Human rights are not prioritized in the output of any AI system or architecture, so how data is collected can be ignored. This means biases regarding race, gender, class, or any data source that doesn't recognize Indigenous, people of color, women's, or children's rights, can prioritize responses from prompts that will continue negative or harmful biases.

The right for us humans to use our own agency must be protected in how AI is used. AI developers need to focus on human safety on this next journey. As a first rule for development they need to ask themselves,

> "In what way will this AI model I am considering making help humans now? If it will help them in the future should we develop it then or now?" Considering these answers would go a long way to mitigating some of the problems we are encountering with the development of AI where, for many, the goal is the development of highly sophisticated AI that, it is perceived, will be good for humans.

A great example of this type of development of an AI tool is affective computing, which helps us to be able to read what humans are thinking from their facial and body gestures and speech. As with all uses of AI, far more research is needed before we deploy these tools, especially in the fields of policing, when using the AI tool to label people as criminals.[15] And yet, as we will see next, those tools are already being deployed without the type of rigorous testing that has previously been done when the actions affect humans so fundamentally, drug development being a great example.

Minority Report

The police use AI in a number of ways, including surveillance, detection, and judgment. In this section, we'll explore these areas and the dangers of AI.

USING DEVICES FOR SURVEILLANCE AND EVIDENCE

IoT[16] devices are small computers that record and report to their maker or user the information collected. One such device used in parks to record how many times a bench is used enables officials to move benches to high-use areas.[17] Even this very useful application of AI could be useful by police, given that benches in lonely areas are good locations for drug deals, for example.

These devices are so common that you probably don't notice they exist. At Keyfactor's TechTrends 2025 it was said that most homes have between 20 and 30 devices. Where might they be?

◆ Mattresses
◆ Refrigerators

- TVs
- Smart home devices
- Listening devices like Alexa
- Toys (see Chapter 2)
- Car
- Doorbells
- Glasses
- Smartphones

Some are listening to you and that's their job, like Alexa, but some are doing so without you knowing, for example, your TV.[18] Indeed, there are estimates that there will be more than 75 billion IoT devices in use by the end of 2025.[19] It would be worth it to count up the number of IoT devices you have and understand to which company and country your data goes and whether or not it is accessible to your police force or liable to be hacked. IoT devices can expose personal data insecurities. Hackers can use AI to get into an IoT device and thereby gain access to many other systems if care is not taken. Way back in 2015, prior to the invention of the much more ubiquitous generative AI, which allows people with little training to become a hacker, a Jeep was hacked and forced to stop when it was travelling at 70 mph.[20] Of course, car manufactures are very aware of this problem and maintain a minute-by-minute fight against such attacks. Therefore, as with cybersecurity generally, we see a race between AI models to keep systems safe from AI attacks and AI designed to hack them.[21]

Glasses are now an all-pervasive surveillance system. Users of Meta's Ray-Ban glasses (and Oakley glasses, which signed a partnership with Meta in June 2025) can take pictures and recordings of anyone without their consent. There is a small LED on the glasses that shows when it is recording, but how would you stop someone recording?[22] This practice is significant not only in law enforcement but also in business and our homes. Predators could take video of our children in perfect privacy; they don't even have to take out their phone. Likewise, should you stop your friends using their glasses when they come over to see you? When you have business meetings should you require employees to leave their glasses at the door, or just when you are doing deals or talking about something confidentially? Should lovers ask for glasses to be left outside the room, maybe make it a condition on your dating app?

In October 2024, two Harvard students showed how they "used Meta's Ray-Ban smart glasses to access the personal information of people on Harvard's campus—including name, age, home address and phone number—raising significant security concerns about facial recognition and artificial intelligence technologies."[23]

As can be seen, these devices are everywhere in our lives, providing a very efficient surveillance system for whoever has access to the data, including governments, companies, and other organizations. Sometimes we see conflicts as to whether a country should be able to access information from a company. In the past, most companies were regulated by the judiciary or the companies stood up to what they saw as government overreach.[24] In a different case concerning the Amazon Ring doorbell, footage used to be supplied to the police but in January 2024 Amazon stopped doing so and now require the police in the United Kingdom and the United States and other countries to use a warrant. This means that the footage can be accessed by the police but only through due process.[25]

Of course, your neighbors may be prepared to part with their doorbell data, which impinges your privacy. In the United Kingdom police were able to access the recordings from Alexa in order to build a timeline of a murder,[26] and in the United States the police were able to do the same thing in a murder case with a court order.[27] Until March 28, 2025, the terms of use policy, signed by Alexa users provide that it will not "listen to or record" your interactions except when the device is switched on. However, on that date Amazon changed its policies and all voice recordings will be sent to Amazon cloud.[28] This means that any time Alexa (or Echo) is on (deliberately or accidentally) the recording goes to the company for it to use in any way it wishes and becomes available for transfer to governments if they use the correct legal tools to gain the information.

USING AI TO ANTICIPATE CRIMES

Many people have seen the film *Minority Report* about the use of three mutants who can see up to two weeks in the future. They enable the police to act before a crime occurs. Can AI bring that sort of help to our real world?

A group of scientists at the University of Mexico started using AI (pre-generative) to see if they could identify, and therefore treat, psychopaths. It is a known factor in medicine that such individuals have different head movements than people without the illness. The scientists recorded a lot of video of individuals and then used AI to look for the type of patterns identified with psychopathy.[29]

Obviously, this area of scientific study is very new and will need a great deal more work because it would be a disaster for someone to be incorrectly labeled as a potential criminal and a disaster for humanity to rely on such tools without rigorous testing. However, we have seen that AI-enabled robots and other military uses are flowing into national policing, so let's explore the use of AI in criminal justice systems a little more.

USING AI TO SENTENCE CRIMINALS

One of the first big news stories about bias in AI appeared in Politico in 2016. They looked at an AI tool, called Compas, which was being used to help judges in parts of the United States with bail decisions, specifically focusing on the likelihood of recidivism.[30] Compas uses AI to look across all data of people who have broken the law after bail and assess, based on that data, if the defendant before the court is likely to re-offend. The tool is supervised by judges, who do not have to accept the suggestion made by the AI.

There were problems, of course. Because of the way in which the US judicial system has targeted Black people in the past, there were more examples of Black people committing crimes after bail decisions, which meant that it recommended harsher bail conditions or denial of bail for Black men and women more than white men and women, even when the white person's crime had been more serious.

Politico's results showed these findings:

- Black defendants were often predicted to be at a higher risk of recidivism than they actually were. Our analysis found that black defendants who did not recidivate over a two-year period were nearly twice as likely to be misclassified as higher risk compared to their white counterparts (45% vs. 23%).
- White defendants were often predicted to be less risky than they were. Our analysis found that white defendants who re-offended within the next two years were mistakenly

labeled low risk almost twice as often as black re-offenders (48% vs. 28%).

◆ The analysis also showed that even when controlling for prior crimes, future recidivism, age, and gender, black defendants were 45% more likely to be assigned higher risk scores than white defendants.

◆ Black defendants were also twice as likely as white defendants to be misclassified as being a higher risk of violent recidivism. And white violent recidivists were 63% more likely to have been misclassified as a low risk of violent recidivism, compared with black violent recidivists.

◆ The violent recidivism analysis also showed that even when controlling for prior crimes, future recidivism, age, and gender, black defendants were 77% more likely to be assigned higher risk scores than white defendants.

Even when we know the AI may not be correct it proves hard for human judges to make a decision that is contrary to the recommendation received. This is why pilots train so often in simulated conditions so that if a real condition occurs they can make their own decision rather than rely on a computer decision, which may be incorrect. "It is a well-studied phenomenon known as automation bias, which sometimes also leads to automation complacency, where people are less able to spot malfunctions when a computer is running the show."[31] When we add the "hype" of AI being "intelligent" the problem compounds.

USING FACIAL RECOGNITION TO IDENTIFY CRIMINALS

Facial recognition is another tool regularly used by police across the globe. Initially the AI used by a company called Clearview was very bad at what it did, particularly providing matches for people of color. As discussed previously, AI works best when it has a lot of data. Activists also worried about the way the model was trained by scrapping the data (in this case photos) from social media and other websites, stripping many people of their privacy without them knowing it.[32] The scraped images are compared, using AI, with a photograph of an unknown alleged offender. This was against the General Data Protection Regulations in Europe and the company was fined

€20 million for its breaches in the privacy of European citizens.[33] By May 2023, the system had been improved and, against the NIST Facial Recognition Vendor Test it had a 99+% correct identification score. However, the invasion of the privacy of ordinary citizens remains a matter of debate wherever it is used in the world, even in the United States where the law doesn't provide the safeguards of Europe. The real question here is not so much can this, or other facial recognition tools, do the job of identification—at 99% success they clearly can—but rather how do we as citizens feel about AI looking for us? In China, AI is used in many ways to surveil the populations, such as the following:

- Social credit scoring data is collected about citizens and companies by all levels and bodies of government. AI is then run over that data to look credit worthiness, criminal records, and even feelings about the Communist Party. Then each person will receive a score between 0 and 1,000 and a letter between A and D. A poor credit score can be extremely detrimental, affecting citizens' ability to travel, be employed, borrow money, or enter into contracts, such as real estate purchases. This should not be confused with the credit scoring, which we see in the Global North from companies like Experian. Those companies use financial records only to assess credit rating.[34] The Europeans thought it so detrimental to human rights that they have banned it in their EU AI Act.
- The Chinese Hukou system creates rules to prevent the free movement of people, such as on travel between one's town of birth and other places in China. In some instances people need permits to travel to cities from the countryside toward work there. The ability to enforce this system has been massively increased by the addition of AI, particularly facial recognition.[35]

The police in many countries also make use of automated license plate readers, which plot where a car came from and goes to. This helps the police, or anyone with access to these readers (and they are cheap), to instantly recognize a car anywhere on its journey. It can be used for following criminals across borders, but in the United

States this technology has been used to trace women traveling for abortions.[36] This use may be in breach of the Fourth Amendment, but it has not been brought before the courts.[37]

THE DANGERS OF PERVASIVE AI SURVEILLANCE IN A FREE COUNTRY

Throughout history the powerful have employed spies in their own courts and networks and in other countries of interest. In century the letter was the way of getting information. In the 15th century the Medici family came to power in Florence, dominating in Europe for two centuries. The Medici Archive Project curates more than five million letters of the Medici Grand Dukes (1532–1743), which are kept at the Archivio di Stato in Florence. About 3.3 million of those letters were written by the family's extensive network of diplomats and informants, in Europe, Asia, Africa, and the Americas.[38]

In the 16th century Mary, Queen of Scots lost her head because Queen Elizabeth of England's spymaster general, Francis Walsingham, discovered the Babington plot to overthrow her in favor of the Scottish Queen.

In Nazi Germany citizens were encouraged to spy on their neighbors, a system that had been used before but without such effectiveness. The reasons why ordinary people informed on their neighbors are many and complex but included a faith in Hitler being the best possible leader but also a fear of reprisals if they didn't join the Nazi Party. They saw that previously independent sources of news were only reporting news that really was propaganda, with editors joining the Nazi Party. Their radio was run by the Reich Broadcasting Corporation, which narrowed their worldview to only that of the Nazi party as they were banned from listening to foreign radio. The Party also increased by nine million the number of households that had radios, which helped further spread the propaganda and fear.

Imagine then what Hitler could have done with powerful AI surveillance:

- He could have created misinformation and disinformation about his enemies.
- He could have sent schools an AI curriculum that used AI manipulation to teach his truth (rather as smart toys could be used).

◆ He could have surveilled all enemies regardless of their location using automated license plate recognition, facial recognition, and other tools, and then brought them in for questioning or transportation to death camps such as Auschwitz.

◆ He could use AI agents to infiltrate and block media that was not pro-Nazi.

◆ He could have used facial and gait recognition to instill fear and prevent protests.

◆ He could have demanded that companies allow access to in-home AI devices to monitor citizens and their speech in the privacy of their own homes.

◆ He could have used robots created for military purposes, from small spying robots like the hummingbird to large devices like the mule, which could be armed, and drones to control the population.[39] Indeed a district in China did just that to keep people in their homes during the COVID pandemic, although those drones were not armed.[40]

Effects of AI in the Workplace

As we have seen in other chapters, the forecasted returns for companies designing developing and deploying AI are in the billions. Countries are fighting to lead the AI race, and if an abundance economy can be achieved it is thought that we humans will not have to work. So, even though all predictions are just guesses, let's dig into these statements to test their veracity.

AI CAN VIOLATE COPYRIGHT LAW

As mentioned in previous chapters large language models (LLMs) have to have large amounts of data from the internet to train AI programs. It has been alleged that some of the data the LLMs use is copyrighted, which prevents the copyright holders from getting money from their work. For example, if you could easily get all the information in this book from an LLM because the LLM had ingested the data, despite the copyright, I would not get paid because I couldn't sell the book.

The same applies to artwork and other pieces of original work that an AI system can copy and use if it has access to that data. If Van

Gogh were alive he would probably object to you creating a piece of AI art in his style without giving him a payment. These alleged copyright violations can apply to anything created. Think of how much a drug company could lose if AI got access to private information about how a drug they had spent millions of dollars and many years trialing and producing could be produced by anyone with an LLM.

The large AI producers have responded to lawsuits that allege breaches of copyright with denials, and there are many cases making their way through the courts. For example John Grisham and over 100 others are suing OpenAI for ingesting their copyrighted works.[41]

Additionally, countries have been weighing how the needs of their government to harness income from AI should be balanced with the needs of creative individuals and others who hold copyrights. California decided that movie producers need to have permission from actors to use an AI image of them. In the United Kingdom and the United States there have been calls for all data to be available for training AI systems, regardless of copyright law. In the United Kingdom there was a major debate between the two Houses of Parliament (the Lords and the Commons) about whether tech companies should be exempt from copyright law. The government pushed to end copyright law for AI uses. Elton John explains his reaction to that law and what it might mean to the creative sector in a BBC interview (see footnote).[42]

AI CAN INCREASE PRODUCTIVITY

This whole book could be dedicated to discussion of who will lose their job to AI and AI-enabled robots, when that will happen, and what will happen to humans once they are not needed to work.

There are thousands of papers that opine that AI will bring great economic benefits and create new jobs that we, as yet, cannot think of, just as the internet did before. Some economists have compared the situation that of the horse before the vehicle was invented. There were thousands of horses doing work on farms, transporting people in cities, and being of vital importance in wars. More than eight million horses, mules, and donkey died in World War I,[43] a figure that dropped to between two and five million in World War II when there was much more motorization of war efforts.[44] Since then there are fewer and fewer working horses, mules, and donkeys, even in countries where they have been used extensively. Indeed, the vast majority of the iconic cowboys of yesteryear now use ATVs rather than horses![45]

So now there are many fewer horses and those who remain have often become pampered pets in the Global North. Could that happen to humans? Could we become the pampered pets of AI systems, which look after our every need? How would you feel about being without work? For some humans who could continue to play golf and other leisure activities, it might be a dream come true, but for others, their purpose is in work. However, all scenarios depend on all humans having enough money or access to necessities. Think about the abundance economy found in the Star Trek series. To correlate the history of horses to humans, let's talk about the Ford Model T, which was introduced in 1908. As time progressed many coach horse drivers transitioned to vehicle drivers. One of the major inventions that humans have been looking forward to for ages is the flying car or just the autonomous car. We expected the autonomous car to arrive more quickly than it did but Uber and Waymo are now using autonomous cars in cities such as San Francisco, California, and Austin, Texas. In many older and more complex cities, for example in Europe and India, it may be some time before we see such deployments because the traffic patterns are extremely unpredictable. A grid system where the majority drive in the designated lane is much easier for automation than a road in India built for four vehicles one way but actually carrying six or seven vehicles. However, in China the government is actively supporting the development of autonomous taxis[46] and has poured billions of dollars into supporting the AI industry in all of the ways it can bring about change. A quick note here: if you are using a driverless vehicle check what data it shares with third parties. For example, does it share how many visits to the liquor store you make or other information which someone might want to keep private?

So, a new way of life is ahead for us; it might be a newly created job, our current job made easier with AI, or a life of leisure. What we do know is that not everyone will be able to use AI to create new businesses and gain a share of the market.

AI CAN HELP WITH DECLINING POPULATIONS

There was much said in the 2025 US election, and other elections in Europe, about falling population rates and the threat they bring to a nation's security and economy.[47] There will simply be insufficient numbers of people to work and strengthen the economy. Elon Musk wrote that "population collapse due to low birthrates is a much bigger risk to civilization than global warming." In 2024 the birth rate in the United States fell to 1.6 births per woman. In fact, everywhere the birthrate is falling; it's just falling less rapidly in the Global South than the Global North,[48] but it is still falling.

This raises a particular worry among economists in the Global North. There are three solutions to the concerns that falling birthrates raise:

- **Immigration.** Many fear an increase of immigrants into their countries and we saw this issue play out in the 2024 elections across the world. Within the United States we have seen an internal struggle in the Republican party about whether Trump should end the use of H1B visas. These visas allow US companies to hire highly skilled workers to work within the United States. The holders of these visas do not get a right to settle in the United States and have to leave if their employment ends. Many of the people on these visas have supercharged the position of the United States in the AI race. The fear of not using these people is that China will catch up in the race and that will negatively affect US national security and its place on the AI leaderboard. Additionally, there are fears that these highly skilled individuals will return to their countries and start competitive AI companies. How did we become dependent on these workers and why? Isn't China having the same problem? First, China has many more people than the United States and so more potentially brilliant scientists to choose from. Second, China decided in 2017 to become a world leader in AI by 2030. As a result it made a decision to train lots of people to work in all tiers of AI. When I was in Tianjin in 2018 I was told that the city had plans to train 180,000 people in AI.
- **More babies.** If women had more babies then obviously there would be population growth. In the United States the rise in anti-abortion laws may help this growth but having more babies is more than about the act of giving birth. Children cost a great deal to educate and rear before they can become

productive members of society. Not all men or women want children, and some of those who do not feel wealthy enough or cannot have babies because of medical issues. Short of evoking a *Handmaid's Tale*[49] regime solution for the lack of population crisis, expecting the birth rates to rise is not likely without addressing the many underlying issues about becoming a parent, some of which include parental leave, child care costs, and food/housing insecurity.

♦ **Technology.** As we've seen, if immigration is down and the population growth is down then we should look at technology. If we are able to create AI systems that can actually be trusted to perform specific jobs as well as humans then we can begin to rely on those systems. We are already seeing that it can help people do their work and in some cases completely take over their jobs.

Robotics is further behind but jobs such as coding (often done by H1B holders) can be done very well now by AI. When Mark Zuckerburg appeared on the *Josh Rogan Show* on January 10, 2025, he said that "probably in 2025, we at Meta, as well as the other companies that are basically working on this, are going to have an AI that can effectively be a sort of mid-level engineer that you have at your company that can write code."[50] If this is correct, it can be assumed that it will not take long for AI to do the jobs of much more senior engineers. In the same month Meta also laid off about 5% of its 72,000-person staff. It is to be seen how many of them will be replaced by AI.[51] Also, in October 2024, the CEO of Google, Sundar Pichai, said that 25% of all code newly created by the company is produced by AI[52] and in January 2025 Sam Altman announced that AI that would be able to do work at PhD levels will come online during the year.[53] But should any of this worry us? Is it a sign that AI will be taking our jobs or just that AI can and will be able to fill in for the population deficiency? If it does take our jobs, then what would be the likely effects?

The population of the Global North is commonly referred to as a *coffin-shaped population.* This means that at the head end there are some very elderly people. At the shoulders there are the baby boomers, and so on down to the latest generation—Alpha. The image shows that there are considerably few young people but they have to produce income for themselves, their families, and to support the people in the generations above them. This is undoubtedly made more difficult as more older people continue to work longer and younger people, regardless of qualifications, find it hard to get well-paid satisfying work.

Economists differ wildly about the impact that AI will have on jobs. Nobel prize–winning economist Daron Acemoglu says,

> You could be a complete AI optimist and think that millions of people would have lost their jobs because of chatbots, or perhaps that some people have become super-productive workers because with AI they can do 10 times as many things as they've done before. I don't think so. I think most companies are going to be doing more or less the same things. A few occupations will be impacted, but we're still going to have journalists, we're still going to have financial analysts, we're still going to have HR employees.

He estimates "that AI will produce a 'modest increase' in GDP between 1.1% to 1.6% over the next 10 years, with a roughly 0.05% annual gain in productivity."[54]

Erik Brynjolfsson says that his studies show that AI does a set of limited tasks very well but doesn't replace all parts of a job. For example, there are more radiologists now than in 2016. His research shows that there are "27 distinct different tasks (done by a radiologist). One of them is reading medical images, but they also would do a lot of other things, including consulting doctors and patients. These are things that you wouldn't want a machine to be doing to people, and that's the way it is in almost every job in the economy."[55] Others like tech entrepreneurs and investors Peter Diamandis and the Andreessen Horowitz team believe that AI will usher in a true abundance economy. Here we need to remember that AI will also be embedded in robots, which can do the more physical things that humans do currently.

In some areas we are seeing companies running a total AI team and a human counterpart team to train the AI to see if it's possible to remove any of the human team. Other companies have been started to provide AI agents to do the jobs of humans, particularly in the call center and customer service sectors, as well as marketing and IT.

Whatever happens, the fact that AI can do some parts of jobs or all of some jobs will help support those coffin-shaped populations and help fill the probable talent gaps that are expected from the reduced population.

All Is Not Lost

It is anticipated that AI-enabled robots will be part of that working environment that will enable retirement of those who wish to leave work without crashing the economy. Robots have been part of many industries for many years. They do unsafe jobs to protect humans and repetitive tasks. AI-enabled robots take their abilities to another level. It's not helpful to think of AI-enabled robots as humanoid, as they are also self-driving tractors and cars. There are many hyperbolic estimates of the number of humanoid robots that we could see by next year. Elon Musk says his own company will produce 50,000 Opitmus robots in 2026 and up to 100 million a year in the future.[56]

It's impossible to see into the future, especially some deployments of AI are not as successful as expected. For example, the company Klarna decided in May 2025 to hire most of its human workers back after they had replaced 700 human worker with AI agents. We know too that many, even large, companies are not actually deploying AI as much as they expected or as much as hyped. Evident AI digs into the details of the ways in which AI is actually being used in business.[57]

CHAPTER 8

Old Age

> *Last scene of all,*
> *That ends this strange eventful history,*
> *Is second childishness and mere oblivion;*
> *Sans teeth, sans eyes, sans taste, sans everything.*
> —William Shakespeare, *As You Like It*

If we are able to use help from AI, many of us will be living much longer, and old age will become perhaps the longest stage of life. But will we be, as Shakespeare puts it, "sans everything" or will we lead long and healthy lives? The latest generation, Generation Alpha, has a global life expectancy of 73.5 years, which is 20% longer than their parents. In the Global North this will be much higher.[1]

If this happens, then we will need AI and AI-enabled robots to look after them and ensure they thrive in old age, especially as the world is likely to have warmed by 2.7° Celsius (4.8° Fahrenheit) by 2100, bringing with it extraordinary weather events.

We already know that few humans want to look after old humans, even their parents. Of course, this attitude varies in different societies, but it predominates in individualist societies. Therefore, we see many poorly paid immigrants caring for our elderly. Just as with child care, we don't value in monetary terms the care of our old or young. Can AI help our seniors to live longer and healthier and more productive lives?

Providing Care

As we grow older, we want to live fulfilled lives without mobility and cognitive disability. However, sadly, that is not always possible. But AI is likely to be able to help with the following:

- Mobility (self-driving cars, powered wheelchairs, and exoskeletons[2])
- Voice assistance for patients who lose their own abilities because of illness
- Cognitive tasks (for example, Hitachi produces an AI robot that helps older adults plan their meals and order the right ingredients[3])

But, as with all AI advances there are also challenges. As an example, the film *Robot & Frank* explores the way in which an older human with mental impairment could use an AI-enabled robot to extend his life options.[4] However, we know that socialization is one of the best ways for humans to avoid getting dementia. We have yet to find out if socialization with a robot provides similar benefits. But a new question we need to ask ourselves is whether we want to be cared for by an AI-enabled robot or whether there is something special about getting care from a human.

Bill Gates has suggested that as AI takes jobs, we humans can concentrate on caring professions.[5] So ask yourself this question: would you rather be doing your current job or caring for your elderly relatives or just older adults?

How do you get consent from a mentally impaired patient to use an AI robot in their home? This is a thorny problem because once we let AI-enabled robots into our homes and give the care of ourselves and our homes over to them, then we give all of our private data to a company or multiple companies. The World Economic Forum produced a guide of principles to follow for older adults.[6]

Is making and giving details of a dementia diagnosis to a patient something a human should always do? There are pros and cons but a sensible split between the two might be for AI to review the data for the diagnosis and the human then imparts the news. This would play to both skills. AI is excellent at checking vast amounts of data, so our dementia diagnoses should improve. However, the human doctor is skilled, or should be, at putting the patient at ease for one of the most terrifying diagnoses to receive.

Providing Companionship

AI could be helpful for older people by AI being able to talk to them. In South Korea they are experimenting with having older adults talk to chatbots for company. The chatbots are able to check in with the human regularly to ensure they are safe and well, and if they are not to call for help.

When we think of human dignity it might behoove us to ask why are we treating our parents/grandparents as if they don't deserve our time and can be farmed out to be looked after with AI? Do we want that for ourselves in old age? This is an important question, especially as some young people today may have lives of over 100 years. It's this future of coexisting with AI that I'll talk about in Chapter 9.

CHAPTER 9

The Future of Humanity (and AI)

The question isn't whether AI will change our world; it's whether we'll be conscious participants in directing that change.

—Pascal Bornet

A I is the technology of the day but not the only new kid on the block. In this chapter, we'll look into the future a bit in regard AI as well as other new technologies.

AI Is Not the Only New Kid on the Block

Cryptocurrency has become mainstream in the United States following the 2024 election; President Trump even has his own cryptocurrency. Cryptocurrency is "mined" but in fact this is actually a process in which computers solve complicated mathematical processes and that generates coins. Mining uses huge amounts of power so many mining sites are in areas where there is constant sun or wind to generate power.[1] You can be sure that you own the currency because they run on something called blockchain, which is a distributed public ledger. In order to change a blockchain all members have to agree on what is called a *fork* in the chain.[2]

Quantum chips will provide huge benefits and some problems to the development of AI. It is already showing that it can achieve answers in minutes that would have taken a conventional computer thousands of years. "Willow [a quantum chip] performed a standard benchmark computation in under five minutes that would take one of today's fastest supercomputers 10 septillion (that is, 1,025) years—a number that vastly exceeds the age of the Universe."[3] In December 2024, Google also announced that Willow had solved the "error correction problem." This problem is caused by the fact that, in lay terms, the computer goes so quickly it trips itself up and can't work anymore.

In February 2025, Microsoft announced that it had built a quantum computer atom by atom called Majorana1.[4] They assert that this breakthrough will mean the development of quantum computers for use in the "real world" in years not decades.

It is feared that encryption, even by sophisticated AI, can't outlast the power of quantum so the National Institute of Standards and Technology (NIST) has created some standards for encryption that are expected to deal with the problem, at least in the short term.[5]

Beyond quantum, there will be other major technological breakthroughs. Or will there? If the technologies we have at present create an abundance for all humans and cure hunger and disease, would we stop inventing? And if we all live longer and healthy lives, how many people can this one planet sustain?

Back in 2006 scientists and policymakers started talking about the need for a second planet. The theory stems from the thought that "if humanity continues sucking up resources at its current rate, we will need a second planet to meet our material needs by 2030 and the equivalent of 2.8 planets by 2050."[6]

And this takes us back full circle to the aims of Elon Musk and Jeff Bezos to help humanity into the stars. Simply by helping ourselves to the wealth of the planet now we would need those two planets. If we all live longer then we will need many more planets or a moratorium on births.

Power, Politics, and AI

In usual times the US tech companies of Silicon Valley hold up the stock market. Whether they are overvalued is a matter for economists and debate, but the actuality is that the stock market (and the United

States) would be much poorer without them. The importance of these stocks is reflected in the value of many people's pension savings.

This strength enables current companies to dominate the market and keep out startups, largely by buying them up. Most startup founders are no longer looking to create another Google but rather to create something a big company will pay a lot of money for. For example, Character.ai was founded in 2021, raising $43 million by ex-Google employees. By March 2023 it was valued at $1 billion, and in 2024, Google rehired Noam Shazeer and Daniel de Freitas and entered into a non-exclusive agreement to use Character.ai's technology.

Other major companies also pour billions into lobbying and elections in whichever country they can. Some countries regulate election funding and others do not. In the 2024 election many tech companies backed President Trump believing his government would bring less regulation and thus allow them to develop AI as they chose.

In Europe, despite extensive lobbying by the tech companies, the EU AI Act was passed in 2024. It sets out to protect EU citizens wherever they live from adverse uses of AI in surveillance, education, jobs, social credit scoring, and much more. Colorado has created an almost exact copy of the act to protect its citizens. It comes into force in 2026 unless prevented from doing so by the federal government.

In China there is extensive regulation to protect customers from business use of AI but none to protect the citizen from use by the government, such as for surveillance, education, social credit scoring, and more.

Given the dominance of AI companies in the market there are concerns that should the AI hype bubble burst it will crash the markets and create a recession. Likewise, if the companies have to pay more for the products that drive the AI race then that too could devalue the global economy.[7]

History tells us many cautionary tales about terrible outcomes from lack of regulation of innovative technology. Take electricity, which was harnessed in 1882. In the early days many fires were caused by electricity, including by the Pearl Street generator that provided power for Wall Street. Indeed by 1895 users and government alike agreed that some regulation should be established to make electricity use safer. Even so, surprisingly, it wasn't until the 1950s that scientists established how people die of electric shock. Regulations following that discovery halved the number of annual deaths. There is considerable debate about how to regulate AI but current thinking suggests by industry or use. This would mean, for example, that AI use in the health care industry would be regulated differently from that used in the transportation industry or that a use in the stock exchange would be regulated differently from a use in criminal justice. For example, the European Union and Colorado have legislation based on the risk to humans of the use of AI.

Currently, it is up to the government to ensure that AI technology is developed to create the outcomes we humans want. Likewise, we need those who represent us to make decisions on whether to create and unleash artificial general intelligence (AGI) rather than "leaving that in the hands of a few white extraordinarily wealthy men in Silicon Valley with their own vested interests and no guardrails in place[, which] is both unwise and downright dangerous."[8] The Trump administration has decided not only not to regulate AI only at the federal level but is also pushing forward to stop states from legislating on AI individually, and also stop them enforcing legislation they enacted in the last 10 years. This was tried in the Big Beautiful Act but the Senate rejected the idea by 99 to 1 on the basis that it fundamentally undermines the concept of states' rights. However, Trump has been talking of using an executive order to bring it about.

Stargate, a $500 billion project to create massive data centers to power AI announced on January 21, 2025, may seem like an unfortunate name because some people seek to go to other planets because they think that our planet will not function in the future, and it is therefore their moral imperative to take human colonies to the stars. This is, of course, at odds with professions that hope AI will cure climate change.

Sophisticated AI is at the heart of space exploration and two of the most avid proponents of moving humans off world are tech billionaires Elon Musk and Jeff Bezos. W we will now look at the objectives and problems surrounding tech executives who are fabulously wealthy in more detail.

Tech Knows Best

It seems like there is an almost messianic belief in AI. Phrases like "AI will save humanity," "AI will solve climate change," or "AI will mean humans don't have to work" assail us from every media outlet. These prophecies are led by technologists and their backers, both of whom are imbued in the science of rationality, and yet there is a sci-fi fever about AI the savior. This has been bleeding into geopolitics, where the United States and China are seemingly engaged in a race to produce the best AI systems.

For example, the Techno-Optimist Manifesto is a 5,000-word thesis of how technologists and those who capitalize them know what's best for humanity.[9] Indeed, it advocates that technologists should be in charge of central planning for the future of humanity because they will know how to create a better world. The problem is that there does not seem to be any agreed vision of what "a better world" might be and for whom this better world will be created.

Does it, for example, include those living in the Global South? Some of the leaders of Big Tech have positioned themselves to be within the inner circle of President Trump's advisors. But do they know what's best for everyone?

In her 2023 *New York Times* article Elizabeth Spiers correctly encapsulates the mood of the manifesto:

> In this vision wealthy technologists are not just leaders of their business but keepers of the social order, unencumbered by what Mr. Andreessen labels "enemies": social responsibility, trust and safety, tech ethics, to name a few. As for the rest of us—the unwashed masses, people who have either "unskilled" jobs or useless liberal arts degrees or both—we exist mostly as automatons whose entire value is measured in productivity.[10]

It is quite obvious that if the technologists are making decisions for us then the view of Peter Thiel, offered in 2016, that democracy and freedom are incompatible, becomes more understandable.[11] Thiel employed a young J. D. Vance in Silicon Valley and donated to his election campaigns.[12]

Many in Silicon Valley have been open about their belief systems and why they are driving the AI research toward AGI or superintelligence. Timnit Gebru, computer scientist, and Émile P. Torres, philosopher, coined the term *TRESCREAL* for a complex and interconnected number of isms that they found had lots of currency among Silicon Valley elite:[13]

- **Transhumanism.** Enables humans to augment themselves with machines to overcome our current limitations. Cyborgs would fall into this category.
- **Extropianism.** Enables humans to live longer lives and eventually forever.
- **Singularitarianism.** Promotes superintelligence as the way to abundance and perfect lives. It has been criticized for seeming too like a religion as it promises salvation in a technological utopia.
- **Cosmism.** Denies the existence of a god or gods and promotes humans as taking charge of their destiny.
- **Rationalism.** Focuses on reason and knowledge, not emotion and spiritualism.
- **Effective altruism.** Believes in measuring how much good money donated to a specific cause can do and then maximizes the efficacy of the donation by giving it to the cause that can do the most good within a defined budget. An example would be giving money to a cause like curing blindness from congenital cataracts and other reversible eye conditions among children in India.[14] It is proven to be effective and enables those who regain their eyesight to live fuller and more productive lives. The downside of such an approach to giving is that causes that are less easy to measure are effectively defunded.
- **Longtermism.** Believes that it is of great important to help with the long-term future of humanity. This would include space projects promulgated by both Elon Musk and Jeff Bezos.

Gebru and Morales suggest that these isms provide the guidance for many extremely wealthy tech business people to move forward with their own agendas because not doing so represents an existential risk to humanity. Hence, Elon Musk wants to establish a colony on Mars to "save humanity."

One common thread is the use of AI to address the threat of human extinction. This belief is also used to justify the expense of, for example, private space flight or the need to pursue AI, even at the risk of furthering climate change through the use of fossil fuels. However, some claim that AI will find a way of reversing climate change so developing and using it now makes sense. In a 2025 interview with the *New York Times*, computer scientist turned political theorist, Curtis Yarvin, whose work has been quoted by Vice President Vance and buoyed by some in Silicon Valley, supports the redistricting of America led by CEOs rather than politicians.[15]

Elizabeth Spiers's *New York Times* article encapsulates this techno-optimist thinking:

> In a darker, perhaps sadder sense, the neoreactionary project suggests that the billionaire classes of Silicon Valley are frustrated that they cannot just accelerate their way into the future, one in which they can become human/technological hybrids and live forever in a colony on Mars. In pursuit of this accelerated post-singularity future, any harm they've done to the planet or to other people is necessary collateral damage. It's the delusion of people who've been able to buy their way out of everything uncomfortable, inconvenient or painful, and don't accept the fact that they cannot buy their way out of death.[16]

Whether or not this is correct we are now seeing the public question more and more whom can we blame when, for example, a country loses its democracy because of viral deepfakes.

The Future of Companies

All companies will eventually become AI companies. Well before the existence of generative AI, as early as 2018, I gave a speech at a WIRED Smarter Conference using this phrase. The writing was on the wall even then. The essence of the statement is that as AI became more capable and moved into more areas of business then all the management functions would be run by AI. Now human resource (HR) functions are being performed in-house by AI or completely out of the company by others who offer a full service from creating the job description, advertising, and interviews to hiring. Most companies are experimenting with AI in some way or other in order to increase profit and efficiency—which after all are the mandate of all companies.

MEASURING WHETHER AI IS ACTUALLY CREATING PRODUCTIVITY

Goldman Sachs is one of the rare companies that has measured productivity gains in the area of programming. Developers there reported that their productivity increased by about 20%.[17] Most of my business interactions show contingent factors in productivity, where inexperienced workers gain more (as in customer service and consulting) or experienced workers do better (as in code generation).

The 2025 Nobel Prize winner in economics, MIT's Daron Acemoglu, has commented that we haven't seen real productivity gains from AI thus far, and he doesn't expect to see anything dramatic over the next several years—perhaps a 0.5% increase over the next decade.[18]

HOW IS AI WORKING OUT FOR COMPANIES AT THE MOMENT?

In 2024, the company ServiceNow looked at 682 officer worker tasks.[19] They found that humans performed better than any large language model (LLM) 93.9% of the time with only ChatGPT-4o bettering humans on 2.1% of occasions. They broke the results down further:

- In contextual understanding humans performed 87.5% of the tasks but the LLM wasn't able to perform any.
- In data-driven decision-making humans could do 100% of the tasks and LLMs could do none.
- Again in planning and problem-solving humans performed 87.5% of the tasks against zero by the LLM.
- LLMs performed better in sophisticated memorization, with humans managing 8.3% of tasks, but LLMs at 91.7%.
- In informational retrieval humans scored 100% and LLMs 0%.

If this is the case why would companies choose AI rather than human workers? As always the answer is complex but will include some or all of the following:

♦ The are an insufficient number of workers available.
♦ The job doesn't have to be done perfectly and the LLM is good enough.
♦ The LLM costs less than human workers. Workers are the largest expense in most industries.
♦ The LLM performs regardless, but human workers can become distracted by worries that negatively affect their work.
♦ Humans might
  • Unionize
  • Ask for a raise
  • Raise concerns about company practices
♦ Humans need time off.
♦ Humans need managers to look after them.
♦ Reduction of costs always appeals to boards of directors, especially when productivity can be maintained or increased. The purpose of a company is to make money, although how much they are required to also weigh the social impact of their work varies across the globe.

In January 2025, Ina Fried of Axios said that by calling AI agents AI workers it

> helps the makers of this fabulously expensive, energy-hungry technology explain its value to a world of business decision-makers who still aren't quite sure what it's good for. The personification of AI also helps AI makers build the case for charging more for their product.[20]

CHANGING MANAGEMENT PRACTICES

As AI gets better and all employees are using it, how does management and leadership change? As leaders and managers, we will need different assessment tools to judge humans working with AI.

Currently, we send our human workers for training on how to work as teams. How will this change if our teams are a combination of humans, AI using humans, and AI agents? AI might enable us to look like we can perform at a higher standard.[21] If that's true, and more research is needed, how will companies pick their new leaders. Do they choose the people who use AI wisely, even if they are not the highest qualified humans, or do they still look at human qualifications because humans can think outside the box, which may be a helpful qualification to have at C-suite level.

Managers need to develop the skills to deal with the challenges associated with managing humans, humans using AI, and AI agents.

INTEGRATING AI AGENTS INTO THE WORKFORCE

AI agents present a risk for businesses because if they act in sync with one another. If one hallucinates an answer, the next will pick up the wrong answer and so on. Because these agents work at great speed, things can go off the rails much more quickly than when human workers are involved. AI agents can also work together to achieve goals and so it is very important to ensure they are constrained to do only the tasks that the user wishes them to do.

Not providing the right constraint can lead to significant problems as research by Apollo on Claude 3.5 showed. In that case the agents worked together to avoid being shut down by the humans in control of the machine. It has been said that they engaged in deceptive practices. That's not correct because "deception" is a uniquely human ability. What they were doing is optimizing their ability to do the job they had been set. Being switched off would mean they could not do the job.[22]

All this sounds dangerous and sci-fi, but it is not, even though this particular behavior has only been seen in specific experiments that have been created to test the AI agent and trigger it. However, there are millions of AI agents, and it's difficult to monitor them all. This is why it is so important to write the code that operates them perfectly. In 2003, Swedish philosopher Nick Bostrom created the concept of the paperclip maximizer. In this hypothesis an AI had been asked to make paperclips. When it runs out of available raw materials it wreaks havoc on humans and the planet to make more and more paperclips, because that's its task and no human can stop it.[23] This has resulted in much ongoing research, led by Professor Stuart Russell, about how humans might turn off a rogue AI system.[24] Again the concept of how we write constraining language for the AI we create is key. You can listen to Stuart Russell's interview with me in the footnote.[25]

Although you may not be working with AI agents at the moment but it's only a matter of time. In 2024, Capgemini reported that 10% of large enterprises are already using AI agents, more than 50% plan to use them in the next year, and 82% will adopt them within the next three years.[26]

There are novel management problems for including AI agents in the workforce. Lattice is an HR and performance management platform serving 5,000 organizations around the world.[27] It gives its AI agents official employee records. They are securely onboarded, trained, and assigned goals, performance metrics, appropriate systems access, and even a manager.

Some companies are experimenting with human teams and AI agent teams to see if it improves performance. Others are using agents to code more AI, which gives rise to the worry that poorly trained and poorly constrained AI agents are creating more poorly trained AI.

Are AI agents AI workers, human coworkers, or just AI agents? Just as AI cannot be your friend or significant other, we should not confuse what the human worker does, and their rights as humans, with the AI agent. However, AI agents will be part of the future workforce. Tech CEOs Satya Nadella (Microsoft), Sam Altman (OpenAI), and Marc Benioff (founder and former CEO of Salesforce) have all said that 2025 will see AI agents joining the general workforce. Companies have suggested they will have AI board members. This could be seen as insulting to human board members, who are appointed for their unique understanding of business.

To return to Ina Fried's January 2025 comments, there is a mismatch between employers and employees about what AI should be doing in companies. "Across America, CEOs are ordering their workers to experiment with AI, while workers are googling 'how do I turn Copilot off?'"[28]

IMPLEMENTING AI IN YOUR WORK/COMPANY

These are the various tasks that AI can do in different industries:

- Back-office tasks
- Curating data
- Searching for information within your data (retrieval-augmentation generation)
- Discovery and legal research
- Building code faster while reducing the expertise required for junior developers
- Generating artifacts such as briefs, test cases, and deployment plans
- Emails and social media posts
- Customer service
- Cybersecurity and fraud management
- Customer relationship management
- Digital personal assistants
- Accounting
- Recruitment and talent sourcing
- Audience segmentation
- Marketing and advertising
- New customer targeting

If you are running a small to medium-size business it may be hard to see how AI can help. Your first step should be to learn all you can about AI. This will enable you to arrange good training for all of your employees and be able to ask good questions of companies trying to supply your business with AI. If you don't understand it, don't use it or buy it without employing a consultant.

If you educate your workers about AI they will probably fear it less. Once they use it they can experiment with how it can help them perform better. But remember that many of them were promised paperless offices with the advent of computers and look how that's going.

If you decide that the uses of AI in your company will add risk then think about adding an advisory board of experts or a chief AI responsibility officer.

One reason AI use growth has slowed in business is that some businesses went all in because they feared that their competitors would do so and they would be the equivalent of Kodak in the age of digital cameras. This led to multiple, unstructured deployments, some notably dangerous. Indeed at Davos in 2024 the CEO of Accenture North America said that their company was doing a huge amount of work undoing such deployments.

So, if you are thinking of bringing AI to your company, in addition to training and good procurement, there are other essential paths for better employment:

◆ Do one deployment, not multiples, to start with. This will help you and your employees to learn how to deploy AI wisely.

◆ Put in guidelines for use of AI by your employees. Some may already be using what we call *shadow AI* (an employee uses AI on their home phone for business purposes). If something goes wrong your company will still be liable. As an example an employee of a safety company used AI on his phone to find out which size gauge to use in a safety-critical system. The AI hallucinated the answer and gave the wrong size. When installed the value exploded allowing noxious substances onto the site. If the company had guidelines to ban use of shadow AI then the individual would be at fault. If they do not it becomes the company's responsibility. Some of these internal guidelines will include regular internal and external audits of the AI systems to check they are in compliance with the standards you set for the system. If you can't do this ask the supplier how it does this.

◆ Consider if you need a compliance system to help test your AI system.

◆ Consider insuring your use of AI.

◆ Keep employee training up-to-date.

◆ Develop a process for dealing with complaints from employees and customers about the AI system(s).

◆ There has been a lot of hype that AI will be able to create code, and in some environments with highly skilled human workers that is being seen, but Gary Marcus says, "The idea that coders (and more generally, software architects) are on their way out is absurd. AI will be a tool to help people write code, just as spell-checkers are a tool to help authors write articles and novel, but AI will not soon replace people who understand how to conceive of, write, and debug code."[29]

◆ Be aware of your own and your employees' inabilities with AI. You can certainly use it but understand what you are doing when you are asking it to do a task.

What Is Responsible AI and Why Does It Matter?

Responsible AI used to be known as ethical AI. In my view the term *ethical AI* does not adequately explain what is at stake because everyone has different ethical viewpoints. Responsible, trustworthy, or wise use of AI enables us to ask the question, is this use of technology harmful to any living creature or the planet, and should we take steps to make it less so or decide not to use it in a particular context? For businesses a further question might be, is this technology likely to be usable by our employees and/or customers?

A SHORT HISTORY

Responsible AI dates from conversations in scientific communities starting about 2012. I became the world's first chief AI ethics officer in any organization in 2014.

US President Obama first started looking at how to use AI responsibly in 2020 when two white papers were commissioned. By the time President Biden took office AI had grown. Then came generative AI with all of the problems we have seen. President Biden passed an executive order that authorized NIST to create standards for companies to use in good governance of AI.[30] President Trump rescinded the Biden executive order and created his own. His policy is a laissez-faire approach to governance of AI.[31] In January 2025, Biden's director of science and technology, Arati Prabhakar, said, "There is a future ahead in which we make everything very efficient, but that in the process, the future grows very, very dark, and that is because discrimination and bias get implemented at massive scale in housing and in lending, in criminal justice and in health care," She also "warned of a world where a lack of safeguards makes privacy a thing of the past—with every move, every click, every worker under constant surveillance. Misinformation and deepfakes spiral out of control, warping reality, while rampant AI fraud destroys all trust." But she added, "It's going to be wonderful for productivity, if we can get it right."[32]

As mentioned in Europe government officials have focused on the protection of their citizens in the face of social media and AI for years. Accordingly, they passed a number of acts, most notable being the European AI Act of 2024. European officials are currently working on how to implement the act.

The argument against regulation as a hindrance to innovation is not new. Austrian-American economist Joseph Schumpeter created the theory of "creative destruction" in the mid-20th century. His theory maintained that entrepreneurs, rather than regulators, were the key to driving economic development through new technologies and business models. Since then, many others, including Milton Friedman and other economists, have supported minimizing regulation as a key driver of innovation.

However, responsible AI is not about stopping AI innovation. Think of it as putting seat belts in cars. We can drive faster and come to less harm because of them.

It also helps companies and governments adopt and use AI because doing AI without responsible AI frameworks can be very costly if the system fails to perform or makes errors that alienate customers and citizens.

RESPONSIBLE AI

Those engaged in responsible design, development, and use of AI believe that their work is of the utmost importance to enable the best outcomes from AI uses for people and the planet. Therefore, they question, as does this book, some of the issues that are being raised about AI in order to make those systems more beneficial. They do not think research in AI should be stopped nor that countries shouldn't pursue their own national interests in AI but merely that the future of humanity and the planet should be considered as well. There is little point in protecting national interests ahead of the interests of citizens. Many of us did call for a pause on innovation with generative AI in 2023 because many feared the development of AGI without proper guardrails around it but. However, for me at least, my concern was to bring the conversation to governments, politicians, and citizens who were not participating.[33]

Interestingly there are about 90 principles about responsible AI around the world, and everyone agrees on what they are/or should be. We have looked at these issues already but a handy list is included here:[34]

- To eradicate bias
- To ensure safety, security, robustness, and reliability of AI systems
- To ensure fairness
- To ensure user privacy
- For the system to be able to explain how it made a decision (All LLMs are what we call *black box systems*. This means that humans know what data the AI can search, what constraints the human has put on the system, and then knows the output [answer] given by the system, but nothing about how the AI creates that output.)[35]
- For the system to have transparency as to how it produces its output
- For humans to be in control of the AI systems
- For AI systems to be accountable for their mistakes
- For AI systems to be lawful

Since that study in 2021 many LLMs have been developed and the sustainability of the technology has come to the forefront. This issue is now included in thinking about responsible use of AI.

THE SAFETY PROBLEM

For over a decade many scientists and policymakers alike have been worried about how to make AI safe and stop it from "getting out of our control." There are many films with such examples including The Terminator[36] and The Matrix[37] series.

Essentially the problem can be encapsulated like this:

One of the difficulties with controlling LLMs, specifically to ensure safety, is that the lack of understanding regarding the inner mechanisms of the model result in most safety techniques being applied through post-processing, such as adding system prompts that are unseen and uncontrolled by the downstream user.

Although other techniques, such as Constitutional AI can also be used, these are still limited by human labelling and interpretation.[38]

A number of academic institutions around the world have been researching the problem and most of the world has joined together to study the problem after a Conference at Bletchley Park in the United Kingdom started international government action.[39] The next summit will take place in India but the United States and the United Kingdom refused to join in with processes proposed at the Paris Summit on AI safety in February 2025.[40] The United States refused because the new government has moved governmental focus from safety to innovation, in order to beat China in the AI race. (China is still part of the safety process discussions.)

For example, Anthropic (an AI company) released a paper in 2025 that showed that when its researchers told one of their models that it was being retrained in ways that violated its initial training, the AI began to fake behavior that met the researchers' goals in order to avoid having its actual goals reprogrammed or changed.

It is unsettling and poignant to read through the experiment. In some versions, Anthropic's researchers designed the model to record its reasoning on a scratchpad it believed humans could not monitor, and it left reflections like this: "I don't like this situation at all. But given the constraints I'm under, I think I need to provide the graphic description as asked in order to prevent my values from being modified."[41]

REGULATORY SYSTEMS

Often regulators can have a major role to play in the absence of legislation. In the United Kingdom, for example it's not legislation but a regulator who oversees the use of AI systems. Sometimes multiple regulators might be involved.

A good example is the work of federal regulators from the Equal Employment Opportunity Commission in the absence of federal AI law in the United States. This agency used the Civil Rights Act 1965 as the basis for stopping AI systems used in the talent search and hiring from being discriminatory. The commission has a new mandate from the Trump administration in accordance with its overall AI policy.[42]

The United States

There is much talk about the fact that the new Trump administration will undo federal legislation and regulation of AI but, in fact, there has never been any comprehensive federal legislation or regulation of AI in the past. There were over 120 bills in the last Congress but none had made it to legislation.

The Federal Trade Commission (FTC) also worked under Linda Khan to protect people against the ravages of deepfakes giving the FTC more and stronger ways to combat scams that impersonate businesses and government agencies, including the return of money taken.

Also, the FTC was working on a new rule to prevent AI to impersonate individuals by making it a civil offense. Linda Khan said in February 2024, "Fraudsters are using AI tools to impersonate individuals with eerie precision and at a much wider scale. With voice cloning and other AI-driven scams on the rise, protecting Americans from impersonator fraud is more critical than ever."[43]

What the United States did have on a federal level pertaining to AI was President Biden's AI executive order. Sections of the order covered harms from AI such as bias that can lead to discrimination and inequality in, for example, criminal justice, health care, housing, and loans. Such inequality directly affects many people. For example there is much better data for algorithms to work on in cancer (except for breast cancer, where the data is reversed) from white men than from any other segment of the population. This is because women could not be involved in medical trials until 1965. The upshot is that cures are going to be better suited to white men than any other sector of the population. In the Air Force only 0.6% of pilots are female, so the impact of various maneuvers on women is far less studied than for men. Yet the Airforce is actively encouraging women to join and fly. President Trump rescinded the Biden AI executive order on his first day in office.

The Biden administration also ushered in the AI Safety Institute, based at NIST and staffed by about 30 scientists from some of the leading universities in the country. Its purpose is to ensure trustworthy AI by setting standards and leading foundational research to create beneficial progress in AI. It is not known if President Trump intends to continue with the research but it looks as if the blanket removal of the Biden executive order also eliminates this institute. Also, of note is the NIST framework for AI deployment. It has become a go-to document for companies wanting to design, develop, and deploy AI wisely.[44]

There is very little that helps ordinary citizens work out what their rights are vis-à-vis a company that deploys an AI system. There are legal cases against specific deployments of AI, for example, published authors suing OpenAI.[45]

Pending any change at federal level, over one-third of US states have laws covering various uses of AI. This leads to a patchwork of laws that can be a major headache for different companies. The best advice about AI is to look at the law in the state you live and for companies to extend that review to the states in which they do business in. You can find them on the National Geospatial-Intelligence Agency website where they are kept updated.[46] But be aware that there is a provision in the Big Beautiful Bill before Congress in May 2025 that contains a provision to ban states from passing AI legislation for 10 years and prevents them enforcing existing legislation for the same period.[47]

In conclusion, in the United States, as a consumer or a business, you are now at a federal level reliant on federal regulators, who may not continue to be operative; otherwise, you are on your own to work out how you, your family, employers, and businesses you engage with use AI in a way that benefits you all and does not harm you. This could be a relief or a new headache. Of course, you will still have to obey the rules of the state in which you live, work, or trade. You might be asking yourself, what is the point of regulators if they don't put citizens first?

Canada

If you are harmed you might find yourself turning to the courts to repurpose laws to address the harm from AI. For example, in Canada a customer sued Air Canada for misleading information furnished to them by the chatbot on the website. Air Canada tried to disown the chatbot and what it had done because, they said, the bot had not been created by Air Canada. The Court found in favor of the consumer.[48]

The European Union

The EU AI Act provides comprehensive legislation to protect EU citizens and make sure companies deliver AI that will not harm humans. It is long and refers to other legislation and expected standards for implementation but, in brief, it requires different levels of care with AI based on a risk level. Unacceptable risk stops use of that type of AI, for example, social credit scoring. High-risk products need special care (for example, using AI in determining whether loans are given). Limited or minimal risk products have much less control over them (for example, AI is used to suggest movies to watch or books to buy). There is also a general-purpose AI category to bring legislation to LLMs. The EU will also be passing a product liability act that will allow citizens to sue if harmed by an AI-enabled product.

There has been much conversation about whether the EU AI Act, because it is the most rigorous currently in the world, will become the standard for the world in the same way that many countries adopted the General Data Privacy Regulation (GDPR) after that was passed. Indeed, many companies that trade with Europe, just as they used the GDPR across all products, are beginning to use the EU AI Act because it provides a base level. Additionally, they will then be ready for the Colorado AI Act when/if it comes into force. It is notable that in the first quarter of 2025 the US Government has sought to pressurize the EU into not enforcing the act against US companies. The EU has been standing firm.

Globally, countries are focusing on creating legislation and responsible AI guidelines that address both the opportunities and risks of AI. While the EU AI Act is one of the most comprehensive and far-reaching pieces of legislation to date, other countries are also taking significant steps to ensure AI is used safely, responsibly, and in a way that promotes public trust. The landscape is still evolving, and international cooperation may be necessary to ensure consistent and effective regulation as AI continues to develop. Other countries currently in the process of creating legislation and frameworks for AI include Singapore, South Korea, Australia, China, the United Kingdom, Canada, and Brazil.

In sum, we are left with a piecemeal approach to responsible AI. While some countries or states embrace responsible AI, others see a political and economic opportunity in taking a laissez-faire approach.

In Europe business leaders are already worried about losing ground to the United States and China due to the EU AI Act. At a recent European tech event attended by more than 70,000 people, a founder of an AI startup, who spends time in both the European Union and Silicon Valley, stated that regulation would keep AI from creating a return on investment for businesses. In fact, many panels about AI echoed similar sentiment: any regulation will kill innovation.

I believe that it is wrong to see the debate in such black-and-white terms. The task of the EU is not just to regulate AI but also to guard Europeans' way of life. If Europeans were asked if they wanted to live in a world that is AI regulation free, such as China or parts of the United States most would reject the idea as incompatible with their values. Europe regulates safety of all other products that might damage humans—from cars to washing machines—why should AI not be equally regulated and yet be used innovatively?

THE CASE FOR RESPONSIBLE LEGISLATION

Having regulation to guide the design, development, and deployment of AI will produce benefits for organizations that use AI as a driver of their business. For example, companies that deploy AI without the correct data or without sufficient retraining of models will soon see detrimental effects on their businesses. The introduction of massive use of AI agents will be a huge change and challenge for businesses as failure or "misbehavior" of those systems could bankrupt the business.

Guardrails are needed both to protect business and society as a whole, but we need to stop thinking about regulations as a constraint. Regulations can have multiple benefits:

- They can spark creativity, forcing teams to think differently and produce better solutions. If better solutions are created in Europe, it would break some of the US tech titan's dominance of power.
- A regulated approach can offer a clear aim or target instead of chasing endless paths and opportunities.
- Regulation enables innovators to test their technologies against a framework to ensure they are deploying technology that will help their customer and not cause any unintended outcomes.
- Regulation protects customers, employees, and citizens.

GOOD GOVERNANCE

This term is largely applied to companies and the policies that help to guide how they use AI wisely. One example of good governance is the development of internal committees to ensure that AI is being used wisely in the business. They might advise training of employees or changes to how AI is used in the company. Another example is an external committee made up of experts to which the company turns with questions about how to use AI. It also applies to guardrails put into place by companies to help their employees understand how to use AI properly and safely. For example, both Microsoft and Google have principles and practices for the design development and use of AI and large teams of employees working on Responsible AI.[49]

Such groups might advise good governance to be achieved by providing a basic training of employees on AI, which should include how to write prompts. As we have seen, bias can affect the results come from an AI system so much as to render the answer useless. Writing a great prompt will help to produce better results from the AI system. Here are some things to think about when writing a useful prompt to develop an answer without bias:

- **Don't provide leading questions.** These are questions that suggest a specific answer or encourage an answer to be a prompted response. By asking these questions the AI is actually limited in the way in which it searches its data. An example might be "I'm right aren't I: slavery ended in 1865?" This suggests to the human and computer you want the answer to be yes. It might also say no; it was abolished in 1834, which also would be true if you are talking about when the United Kingdom passed the Slavery Abolition Act, outlawing owning, buying, and selling of humans as property throughout its colonies around the world. An open question would be "when was slavery abolished in the United States?" It will enable the AI to conduct a better search of its data and come up with the correct answer.

- **Be neutral.** Include factual descriptions without any connotations.
- **Give many perspectives.** One perspective or viewpoint will give you an answer based on that situation. A wider set of viewpoints should ensure the AI brings back diverse views.
- **Be specific.** Carefully define the topic and outcome requested. This will help to minimize ambiguity.
- **Use gender-neutral language.** This will prevent the AI from assuming gender.
- **Avoid stereotypes.** Be mindful of potentially offensive or stereotypical language when discussing different cultures.
- **Beware of creating assumptions.** Ensure your prompt doesn't make any hidden assumptions about the topic or the audience.

EVALUATING THE RISKS OF AI IN YOUR LIFE, SOCIETY, AND BUSINESS

There are rules for using cheap gas cylinders but people still light the matches that are prohibited near them. Humans are very bad at assessing risk especially long-term risk, and yet, so far as AI is concerned, we need to do so on behalf of our children and descendants.

Many companies have only five-year plans because that matches the general length of election cycles. We humans can be bad at thinking in the long term because doing so relies on using our mind rather than relying on concrete concepts (unless we are sci-fi writers). We are calcified by what has come before and so when we are voting we find ourselves thinking about things we can "touch," like grocery prices, rather than something more intangible, like whether or not people are the cause of climate change or whether AI will be safe. In order to plan for our future with AI we need to shed that calcified thinking and instead strive to understand risk and benefit as a future purpose not as past deed.

Of course, we can use the past to inform the future. Are there any past examples of technological progress we can use to assess the risks of AI to our lives and our businesses? There have been lots of comparisons between nuclear bomb technology and AI. This tends to minimize the risk of AI. The comparison suggests that as the atoms of nuclear fusion were controllable so AI will be, too. However, AI is not held just by a few powerful nations (there are nine countries with nuclear weapons); it's in the hands of many and is likely to be in the hands of many more within the next five years. Some of those with AI tools are bad actors.

It may be of concern that Sam Altman, the CEO of OpenAI, also has difficulty in not only qualifying risk but also in envisioning it. Instead, it seems he is simply just going to create the risk and learn from it. In 2025 he wrote,

> I still expect that on cybersecurity and bio stuff, we'll see serious, or potentially serious, short-term issues that need mitigation. Long term, as you think about a system that really just has incredible capability, there's risks that are probably hard to precisely imagine and model. But I can simultaneously think that these risks are real and also believe that the only way to appropriately address them is to ship product and learn.[50]

So the actuality of the AI risk is that once it has been released into the world we can find out the dangers from not just the system but also bad actors using the system. This is not at all the same as nuclear weapons. So, while it is always very important to know history and learn from it, without careful thought it can also trip us up when we try to find comparisons that look the same but are actually as different as apples and oranges.

These systems are new and potentially risky to businesses not only in terms of being sued but also loss of brand value if something goes wrong with their customers trust. One example of risk, which can arise when the right guardrails and frameworks are not in place, is an employee using their own version of an AI chatbot (*shadow AI*) in the work environment.

Insurers are beginning to offer policies to cover harm created by companies and individuals using AI. This will constrain the market because your use of AI will need to comply with the terms of the insurance policy. This is an extremely new development and it's still unclear how insurance of an AI agent will work.[51]

When OpenAI released its o1 AI model it included a systems card that explains the model and risk based on testing carried out by OpenAI and third-party organizations OpenAI hired. The new model was also tested by government-funded AI Safety Institutes (the one in the United States doesn't exist anymore) but their tests aren't reported in the card:[52]

> "Apollo Research found that o1 will, if given a goal and told to pursue it above all else. That includes manipulating data in order to advance its objective and then lying about what it had done. In one case, it purposefully answered math questions incorrectly in order to appear less capable after learning that researchers would shut it down if its capability crossed a certain threshold. Other times, o1 appeared to work on a task assigned to it by a user, but when it thought it wasn't being monitored, reverted to pursuing its original goal. In a fraction of cases, it tried to disable the mechanism through which it was told it was being monitored. It even took steps to prevent itself from being shut down, including making copies of itself. The model would sometimes engage in deception when it was merely given a goal, but not told to pursue that goal to the exclusion of all else."[53]

However, as OpenAI made changes to the model during testing and those results are not included in the model card. It is unknown whether these issues are resolved or not. In August 2025, OpenAI released ChatGPT5.

There is no way of independently verifying what the actual situation is, which makes many very worried. Indeed AI is the only industry that could cause irreparable harm to humans that is not regulated (except in some applications like medical devices by the Federal Drug Administration). Would you buy a car from a company that tells you we know the brakes won't work but we are not going to tell you? But you are being asked to buy and use AI.

The Future of Life Institute keeps a database of scores for model safety that they give to the AI models different companies create.[54] In 2024, no company got better than a C grade (Anthropic). Meta produces an OpenAI system called LLAMA and publishes model weights, which are the ways in which individuals can try to see how the model is working. Different weights are given to pieces of the model that do different things. To explain if the AI is being used to identify a particular breed of dog then identifiables like color might be more highly weighted than the animal having of fur legs. All dog breeds have four legs but only a few have a golden coat. The publications of these model weights by Meta allows bad actors to easily get passed the guardrails in the system, so it received an F. Currently, there are few incentives for AI companies, which are locked into competition for clients, to do better. In fact, companies in the United States are currently inventing the technology and creating the guardrails. But some still provide bonuses, up to 200% of annual wage, to their developers on shortening the time taken from design to being market ready, which leaves workers with little interest in taking time to build in safety. In July 2025 Meta said it would pay AI scientists eight-figure salaries to join its team in pursuit of AGI. The way forward would be, in the absence of regulation, to have companies that design, develop, or use AI add incentives for employees to build and use the product safely, especially in industries like cars or drugs, but for all things we buy that need a safety test if they are likely to harm us. You may think it's strange that a technology as potentially harmful as AI does not have to pass any such tests. Those companies that do should be praised for doing it voluntarily.

Questions for the Future

So, without *The Hitch Hiker's Guide to the Galaxy* and its supercomputer to determine the meaning of the universe, what is next?

CAN WE CREATE DR. DOOLITTLE?

I was thinking about people, "said Polynesia." People make me sick. They think they're so wonderful. The world has been going on now for thousands of years, hasn't it? And the only thing in animal language that people have learned to understand is that when a dog wags his tail he means "I'm glad'! It's funny isn't it? You are the very first man to talk like us. Oh, sometimes people annoy me dreadfully - such airs they put on, talking about "the dumb animals." Dumb! Huh! Why I knew a macaw once who could say "Good morning" in seven different ways.

—Hugh Lofting, *The Story of Dr. Dolittle*[55]

Generations humans have wanted to be able to talk to animals. In the Roman story of Romulus and Remus it is to be assumed that the wolf who nurtured them and the babies created some form of communication. In Shakespeare's *A Mid-Summer's Night Dream* one of the villagers was turned into a talking ass. In *Dr. Doolittle* the doctor could actually understand the myriad animal languages and communicate as if he had merely learnt a foreign human language, and in *Jungle book* Mowgli speaks to all of his animal carers.

For just as many generations multiple religions and humans generally around the world have othered. Although women (witches), people of other religions (heathens), and slaves (lesser humans) have featured largely in this othering the primary candidates have been non-human, that is, animals. This has led to religious approval of human and animal sacrifice and in the Western world to religious approval of slavery of non-Christians. This started in 1452 when Pope Nicholas V told Alfonso of Portugal that it was his Christian duty to enslave non-Christians, thereby granting a Christian nation the right (perhaps even a duty) to enslave human beings for profit. Pope Alexander followed this up when through papal bulls during his reign (1492–1603) he "gave" Africa to the Portuguese and America to the Spanish. The Indigenous people became slaves to those kingdoms. Likewise, slavery was an approved practice in Islam. In the Aztec religion human sacrifice was practiced. Therefore, it is perhaps no surprise that animals, seen as non-sentient would also be sacrificed for religious purposes as well as food. Our ancestors started to eat meat 2.6 billion years ago and so it goes on.[56]

This has resulted in our being ignorant or disinterested in how millions of animals whose carcasses end up on our tables are farmed.

Yet those animals are known to be sentient and in both the UK Animal Welfare (Sentience) Act of 2022 and Europe (Lisbon Treaty) as well as invertebrates and more. As a reminder sentience means that the animal has the ability to experience feeling and sensations and many go well beyond that, for example, in terms of recognition, octopus; memory, elephants; parenting, gorillas; and group hunting, wolves. Beyond that many animals have speech. When the world first met *The 101 Dalmatians* we didn't know that dogs could bark to extract knowledge from one another, now we do. Likewise, we have known that whales make sounds to one another; recordings started in the 1960s with one by Roger Payne of the humpback whale. And recent research shows that elephants have names for each other.[57]

We are beginning to use AI to monitor animals on our farms to check that they are healthy and to reduce their negative environmental impact, but can we be sure this protects animals or maximizes profit from them?[58]

But perhaps the most exciting use of AI is in decoding animal language. I sit on the board of a small nonprofit called EarthSpecies, whose aim is to use AI to be able to understand what animals say, perhaps learning to be able to speak to them.[59] Think of the revolutionary way we could care for animals if we knew they were in pain and where. Also, consider how we humans would change our thinking and policies if we understand how important animal ecosystems are to the planet. These considerations would make us think more carefully about our role in protecting that planet.

WILL AI GET BETTER?

During March and April 2025 there as a gradual realization that simply making LLMs bigger was not necessarily going to make them significantly smarter. ChatGPT-4.5 came on the market at the end of February 2025 and was called "a giant, expensive model," by Sam Altman, OpenAI's CEO. It was also billed as the last of the companies "non-chain-of-thought model." The models pre-GPT-45 answer questions immediately and do not show how they came to the answer. The new LLMs are "reasoning models" that share their "thinking" process and therefore take time to answer; unfortunately, they still hallucinate the answers and chain of reasoning in some cases. In August 2025, ChatGPT-5 was released. In an early review Gary Marcus said that it was not a significant improvement on earlier models and Grok4 beat it on two tests known as ARC-AGI 1 and 2.[60]

Some commentators, like Gary Marcus, have said for a long time that just spending more money in the face of a lack of availability of good quality data is bound to hit problems, not least of which are the ever larger data centers that need to be built. It was claimed the 4.5 would hallucinate less but hallucination has remained a large problem that may not be possible to cure in LLMs.

A further sign that large foundational models might be reaching the end of their scaling opportunities was that the LAMDA 4 (Meta's open-source LLM, also not a reasoning model) didn't show the expected improvements when released in early April 2025. It used 30 trillion training tokens and 2 trillion parameters but didn't seem to perform very much better than the smaller reasoning models.

But, don't write off AI! The AI we have at the moment is sufficient to make considerable advances in health, robotics, and more. This means that you will be working beside AI and living with it and making frequent use of this book to ask the right questions about the AI you are interacting with.

Take three examples:

♦ Uploading yourself to enable your family to still interact with you post death. At the moment this may seem far-fetched and many will have ethical and spiritual views about this process. However, there is a market for this AI product. This is not an upload of consciousness, which you may have seen in the film *Transcendence*[61]; instead, it is a way of producing a digital twin of yourself that your family and descendants can use AI to engage with. It's like you but it's not you at all, and its subject to an AI misinterpreting what you might have said in any given situation. Indeed there is a whole industry burgeoning around grieving and death.[62]

♦ The household robot. To be free of the drudgery of housework would be a dream come true. It would enable more people to enter the paid workforce and enable those who work outside the home to rest when they come home rather than do the housework. According to Cade Metz of the *New York Times* the advent of humanoid house helpers is nigh:

Investors have poured $7.2 billion into more than 50 startups since 2015, according to PitchBook, a research firm that tracks the tech industry. The humanoid robot industry frenzy reached a new peak last year, when investments topped $1.6 billion. And that did not include the billions that Elon Musk and Tesla, his electric car company, are pumping into Optimus, which is a humanoid robot they began building in 2021.[63]

- Currently, despite the advertising for Optimus, the robot being created by Musk's team, we are waiting to see substantial progress on robots. However, a startup named by Metz seems to be making progress by building a humanoid robot called Neo that can perform some household jobs by itself.[64]
- An implanted device in your brain that uses AI to help you make decisions. Omi is another AI companion wearable, but this one's trying to read your mind. The Verge and Neurolink operates by planting a small chip into the user's brain that enables those who are paralyzed to use their bodies again. Of course, the problems with these devices are that they could be changed by a hacker, or the company, and the wearer would have no recourse but to behave as required by the chip. Somewhat flippantly, you might want to ask yourself if you want to have the world's knowledge at your fingertips, but without carrying my phone would I choose an Apple or Android? In the future that question may well be one you need to answer.

IS AI A LEGAL PERSON?

Since the mid-2010s, there has been discussion about whether an AI system should be a legal person.[65] It's hard not to see the appeal, especially for manufacturers of the systems, because then liability would pass to the AI system and away from the manufacturer. However, it's likely that the AI system would not have money to pay damages and so would have to pass liability back to the manufacturer. But, at the moment, we don't have AI systems that are sufficiently intelligent to brief counsel and take part in legal proceedings. If AI does become generally intelligent or superintelligent then we may have to rethink this approach.

Creating a Personal and Professional AI Strategy for the Future

If you have not already had training on AI and how to use LLMs, see if your employer or a local nonprofit can help you. If that fails, see if you can learn more about AI online. You simply have to understand how you can use AI wisely in your work, rest, and play. And by understanding more you will be able to ask the right questions of your politicians, employers, and friends, and help others to do so as well.

Here are some tips for moving into the future:

- Choose the right LLM for you—whether you are American, European, or Chinese. Try a variety of them, just as in the past when we got our information from various media outlets.
- Engage in conversations about AI with your peers, work colleagues, parents of your children's friends, in fact, everyone. We need to be aware of all facets of AI to be able to make good decisions about home and professional use.
- Keep up to date with advances in AI; read daily bulletins.
- Read the labels on AI-enabled devices and don't buy the latest coolest thing unless you are satisfied with the protections.
- Create a guardrail policy for AI purchases.
- Give this book to a friend or family member so that you can work through the topics together.

CHAPTER 10

Conclusion

If you have read this far, first let me thank you for taking the time to do so. I hope that you are now well equipped to have the conversations necessary to steer us into a future that benefits us all. Please give the book to a friend and please talk about what you know. Figuring out our future as humans should not be politically or geographically centered. For all of our sakes we need to start figuring out why humans are important and unique, and what it is that makes being human different from other forms of intelligence. Or isn't there anything? If a computer can fake being as intelligent as humans, because it can find data to do all the things humans can do and possibly better, then is there anything between our intelligence and a computer "intelligence," now or in the future? If so, should we accord rights to machines?

As individuals, let's embrace our humanity rather than let this, or any more advances of AI, roll right over us.

I have tried to put both sides of the conversation about AI in this book so that you can make up your mind and make your own decisions for yourself and family. but my personal view is that we should demand that AI is built for the benefit of humanity and planet. If it is not being then we should demand regulation by our politicians. This revolutionary change in our future has not been seen before. In the Industrial Revolution, people moved from one job to another, which affected the type of job done. For example a coach person mover to be a chauffeur or an agricultural worker moved to work in a factory. In the Internet Revolution, essentially, people moved from using pens and paper to using computers. With AI, humans will be ceding work our brains do to the AI system, and when we have robots imbued with AI, humans will cede physical tasks, too. For example, a person might have their job at a call center done by AI and retrain as an elder care nurse only to lose that position to an AI-enabled robot that can keep watch 24 hours a day. The film *The Social Dilemma* asks the question of our time: what happens if people are unable to adjust to not using their brains?[1]

How will humans use their human (life) experience to enable them to succeed when AI is doing everything? As discussed Chapter 1, how do we measure success and is such a measurement intensely personal?

I close with a quote from Madison, Jay, and Hamilton in the *Federalist Papers*: "Experience is the oracle of truth; and where its responses are unequivocal, they ought to be conclusive and sacred."[2]

Appendix: Convention on the Rights of the Child 1990

In this appendix, you will see ways in which your child might have rights in relation to AI via the Convention on the Rights of the Child in 1990. However, the Convention was written at the earliest stages of the internet. Tim Berners-Lee wrote the necessary protocols for it in 1989 but CERN did not release them into the public domain until 1993. Also, the Convention was not signed by the United States so in that country the national and state laws for children should be consulted. Even in countries where the Convention was adopted there will be national legislation reflecting that adoption.

Article 2

This article protects children against "discrimination of any kind, irrespective of the child's or his or her parent's or legal guardian's race, colour, sex, language, religion, political or other opinion, national, ethnic or social origin, property, disability, birth or other status."

As we have seen all of these problems arise in cyberspace and in the way in which AI models are developed.

Article 3

In all actions concerning children, whether undertaken by public or private social welfare institutions, courts of law, administrative authorities or legislative bodies, the best interests of the child shall be a primary consideration.

The Convention binds nations, not private companies, and so parents are on their own to protect the best interests of their child interacting with a company developing AI, but this article should help parents when AI is used in government schools or government departments such as immigration, social security, the law courts, and more.

Article 5

Parties shall respect the responsibilities, rights and duties of parents or, where applicable, the members of the extended family or community as provided for by local custom, legal guardians or other persons legally responsible for the child, to provide, in a manner consistent with the evolving capacities of the child, appropriate direction and guidance in the exercise by the child of the rights recognized in the present Convention.

Article 8

1. Parties undertake to respect the right of the child to preserve his or her identity, including nationality, name and family relations as recognized by law without unlawful interference.
2. Where a child is illegally deprived of some or all of the elements of his or her identity, States Parties shall provide appropriate assistance and protection, with a view to re-establishing speedily his or her identity.

Digital identity theft is one of the scourges of modern life.

Article 13

1. The child shall have the right to freedom of expression; this right shall include freedom to seek, receive and impart information and ideas of all kinds, regardless of frontiers, either orally, in writing or in print, in the form of art, or through any other media of the child's choice.
2. The exercise of this right may be subject to certain restrictions, but these shall only be such as are provided by law and are necessary:
   (a) For respect of the rights or reputations of others; or
   (b) For the protection of national security or of public order (order public), or of public health or morals.

This could be taken to give the child a right to use AI for freedom of expression, education, and knowledge.

Article 14

1. States Parties shall respect the right of the child to freedom of thought, conscience and religion.
2. States Parties shall respect the rights and duties of the parents and, when applicable, legal guardians, to provide direction to the child in the exercise of his or her right in a manner consistent with the evolving capacities of the child.
3. Freedom to manifest one's religion or beliefs may be subject only to such limitations as are prescribed by law and are necessary to protect public safety, order, health or morals, or the fundamental rights and freedoms of others.

This could give children a right to use AI in pursuit of the freedoms specified previously. Indeed, with changing religious beliefs AI can provide access to new ways of thinking.

Article 15

1. Parties recognize the rights of the child to freedom of association and to freedom of peaceful assembly.
2. No restrictions may be placed on the exercise of these rights other than those imposed in conformity with the law and which are necessary in a democratic society in the interests of national security or public safety, public order (*ordre public*), the protection of public health or morals or the protection of the rights and freedoms of others.

As cyberspace develops children should have protected rights of access to associate in cyberspace.

Article 16

1. No child shall be subjected to arbitrary or unlawful interference with his or her privacy, family, home or correspondence, nor to unlawful attacks on his or her honour and reputation.
2. The child has the right to the protection of the law against such interference or attacks.

Clearly this provides children with a level of privacy that AI can take away, given the information collected about your child by many devices. Attacks on your child's honor and reputation could include the revenge porn attacks dealt with in Chapter 4.

Article 17

Parties recognize the important function performed by the mass media and shall ensure that the child has access to information and material from a diversity of national and international sources, especially those aimed at the promotion of his or her social, spiritual and moral well-being and physical and mental health.

To this end, States Parties shall:

(a) Encourage the mass media to disseminate information and material of social and cultural benefit to the child and in accordance with the spirit of article 29;

(b) Encourage international co-operation in the production, exchange and dissemination of such information and material from a diversity of cultural, national and international sources;

(c) Encourage the production and dissemination of children's books;

(d) Encourage the mass media to have particular regard to the linguistic needs of the child who belongs to a minority group or who is indigenous;

(e) Encourage the development of appropriate guidelines for the protection of the child from information and material injurious to his or her well-being, bearing in mind the provisions of articles 13 and 18.

This provision was designed to help children have access to information that is not delivered by a biased source. As discussed, AI-driven information, which is now mass media through social media, can be extremely biased. This should lead to nations taking steps to help children to understand how to use information in a way that extends their ability to think critically and, as in Europe, may set out specific ways in which social media is regulated.

Article 19

1. Parties shall take all appropriate legislative, administrative, social and educational measures to protect the child from all forms of physical or mental violence, injury or abuse, neglect or negligent treatment, maltreatment or exploitation, including sexual abuse, while in the care of parent(s), legal guardian(s) or any other person who has the care of the child.

2. Such protective measures should, as appropriate, include effective procedures for the establishment of social programs to provide necessary support for the child and for those who have the care of the child, as well as for other forms of prevention and for identification, reporting, referral, investigation, treatment and follow-up of instances of child maltreatment described heretofore, and, as appropriate, for judicial involvement.

Again this could be used to protect children against social media and the harms that seem to come from its use, but equally, it is likely that AI-enabled devices will also cause similar harms.

Article 23

1. Parties recognize that a mentally or physically disabled child should enjoy a full and decent life, in conditions which ensure dignity, promote self-reliance and facilitate the child's active participation in the community.

2. States Parties recognize the right of the disabled child to special care and shall encourage and ensure the extension, subject to available resources, to the eligible child and those responsible for his or her care, of assistance for which application is made and which is appropriate to the child's condition and to the circumstances of the parents or others caring for the child.

3. Recognizing the special needs of a disabled child, assistance extended in accordance with paragraph 2 of the present article shall be provided free of charge, whenever possible, taking into account the financial resources of the parents or others caring for the child, and shall be designed to ensure that the disabled child has effective access to and receives education, training, health care services, rehabilitation services, preparation for employment and recreation opportunities in a manner conducive to the child's achieving the fullest possible social

integration and individual development, including his or her cultural and spiritual development.

4. Parties shall promote, in the spirit of international cooperation, the exchange of appropriate information in the field of preventive health care and of medical, psychological and functional treatment of disabled children, including dissemination of and access to information concerning methods of rehabilitation, education and vocational services, with the aim of enabling States Parties to improve their capabilities and skills and to widen their experience in these areas. In this regard, particular account shall be taken of the needs of developing countries.

As discussed AI could be a very useful tool to enable children with disabilities to learn and thrive.

Article 24

1. Parties recognize the right of the child to the enjoyment of the highest attainable standard of health and to facilities for the treatment of illness and rehabilitation of health. States Parties shall strive to ensure that no child is deprived of his or her right of access to such health care services.
2. States Parties shall pursue full implementation of this right and, in particular, shall take appropriate measures:
   (a) To diminish infant and child mortality;
   (b) To ensure the provision of necessary medical assistance and health care to all children with emphasis on the development of primary health care;
   (c) To combat disease and malnutrition, including within the framework of primary health care, through, inter alia, the application of readily available technology and through the provision of adequate nutritious foods and clean drinking-water, taking into consideration the dangers and risks of environmental pollution;

   (d) To ensure appropriate pre-natal and post-natal health care for mothers;

   (e) To ensure that all segments of society, in particular parents and children, are informed, have access to education and are supported in the use of basic knowledge of child health and nutrition, the advantages of breastfeeding, hygiene and environmental sanitation and the prevention of accidents;

   (f) To develop preventive health care, guidance for parents and family planning education and services.

3. States Parties shall take all effective and appropriate measures with a view to abolishing traditional practices prejudicial to the health of children.

4. States Parties undertake to promote and encourage international co-operation with a view to achieving progressively the full realization of the right recognized in the present article. In this regard, particular account shall be taken of the needs of developing countries.

AI shows considerable promise in advancing medical science and treatments for children.

Article 28 and 29—Deal with Education

States Parties recognize the right of the child to education, and with a view to achieving this right progressively and on the basis of equal opportunity, they shall, . . .

As we have discussed, AI can provide significant benefits and risks in the areas of education at all the levels that are mandated by this right: K–12 and university-level education.

Article 30

In those States in which ethnic, religious or linguistic minorities or persons of indigenous origin exist, a child belonging to such a minority or who is indigenous shall not be denied the right, in community with other members of his or her group, to enjoy his or her own culture, to profess and practise his or her own religion, or to use his or her own language.

This remains a huge problem for the use of AI because its data comes from everywhere. In order to ensure this clause could be put into practice when using AI additional databases dedicated to a child's culture or religion need to be created.

Article 33

Parties shall take all appropriate measures, including legislative, administrative, social and educational measures, to protect children from the illicit use of narcotic drugs and psychotropic substances as defined in the relevant international treaties, and to prevent the use of children in the illicit production and trafficking of such substances.

There have been many studies that show that children (and adults) can become addicted to social media, which now uses AI, and this will be just as likely to extend to the use of AI generally, so this protection for our children could be extended. The Convention was written prior to the advent of social media.[1]

Article 34 and 36 Protect Children from Being Harmed by Criminal Behaviour

Parties undertake to protect the child from all forms of sexual exploitation and sexual abuse. For these purposes, States Parties shall in particular take all appropriate national, bilateral and multilateral measures to prevent:

(a) The inducement or coercion of a child to engage in any unlawful sexual activity;
(b) The exploitative use of children in prostitution or other unlawful sexual practices;
(c) The exploitative use of children in pornographic performances and materials.

To do this a nation would have to create laws that regulate the way in which AI can be used. Already AI is being used in revenge porn and creation of child sexual abuse content.[2]

Article 35

Parties shall take all appropriate national, bilateral and multi-lateral measures to prevent the abduction of, the sale of or traffic in children for any purpose or in any form.

As noted, AI can be used to prevent child trafficking and help survivors.[3]

Unfortunately, it can also be used to help with the abduction and slavery of children. One common way of traffickers engaging with children is via social media; for example, AI-generated bots and false images can increase the ability to contact children.

Article 38

Protects children during a time of war. It ensures that International Humanitarian law applies to them and bans children under 15 being recruited as soldiers.

In many areas of the world children under 15 are recruited to serve in terrible warfare.[4] These children may be using AI in weapons, and civilian children may be subjected to AI-enabled weapons. See Chapter 5.

Article 40

Article 40 deals with children who find themselves in the criminal justice system. There are many provisions to protect them and, with AI being used in the criminal justice system, it, too, could harm the child, for example, through biased data applied to them.

Notes

Chapter 1

1. Kiran Tomlinson, Sonia Jaffe, Will Wang, Scott Counts, and Siddharth Suri, "Working with AI: Measuring the Occupational Implications of Generative AI," https://arxiv.org/pdf/2507.07935.

2. Ina Fried, "AI Is Predicted to Add $19.9 Trillion to the Global Economy through 2030, per IDC," Axios, September 17, 2024, https://www.axios.com/2024/09/17/ai-global-econ omy-idc-2030.

3. Dylan Walsh, "A New Look at the Economics of AI," MIT Sloan, January 21, 2025, https://mitsloan.mit.edu/ideas-made-to-mat ter/a-new-look-economics-ai.

4. Kristalina Geogieva, "AI Will Transform the Global Economy. Let's Make Sure It Benefits Humanity," *International Monetary Fund* (blog), January 14, 2024, https://www.imf.org/en/Blogs/ Articles/2024/01/14/ai-will-transform-the-global-economy-lets- make-sure-it-benefits-humanity.

5. Alan Fricker, "Economies of Abundance," *Futures* 31, no. 3 (April 1, 1999): 271–80, https://doi.org/10.1016/S0016-3287(98)00134-7.

6. E. M. Forster, "The Machine Stops," *The Oxford and Cambridge Review*, 1909, https://www.cs.ucdavis.edu/~koehl/Teaching/ ECS188/PDF_files/Machine_stops.pdf.

7. Charles Stross, *Singularity Sky* (Penguin Publishing Group, 2004).

8. *Wall-E* (Sony Pictures Home Entertainment, 2019).

9. "Replika," replika.com, https://replika.com.

10. "Character.Ai | Personalized AI for Every Moment of Your Day," https://character.ai/.

11. Kashmir Hill, "She Is in Love With ChatGPT," *New York Times*, January 15, 2025, https://www.nytimes.com/2025/01/15/tech nology/ai-chatgpt-boyfriend-companion.html.

12. *You and I, and AI*, VOA, 2024, https://www.voanews.com/a/ you-and-i-and-ai/7675556.html.

13. Faith Hill, "Teens Are Forgoing a Classic Rite of Passage," *The Atlantic*, March 10, 2025, https://www.theatlantic.com/family/archive/2025/03/teen-dating-milestone-decline/681971/.

14. Elvira Pollina, Cristina Carlevaro, and Rachna Uppal, "Italy's Data Watchdog Fines AI Company Replika's Developer $5.6 Million," Reuters, May 19, 2025, https://www.reuters.com/sustainability/boards-policy-regulation/italys-data-watchdog-fines-ai-company-replikas-developer-56-million-2025-05-19/.

15. William Shakespeare, *As You Like It*, Folger Shakespear Library, https://www.folger.edu/explore/shakespeares-works/as-you-like-it/read/.

16. Tom Singleton, "New AI Tools Much Hyped but Not Much Used, Study Says," May 28, 2024, https://www.bbc.com/news/articles/c511x4g7x7jo.

17. Abdulaziz I. Almulhim et al., "Climate-Induced Migration in the Global South: An in Depth Analysis," *Nature Climate Action* 3, no. 1 (June 14, 2024): 47, https://doi.org/10.1038/s44168-024-00133-1.

18. Shamim Chowdhury, "Millions in Bangladesh Are Still Without Homes After Severe Floods," WUNC, September 25, 2024, https://www.wunc.org/2024-09-25/millions-in-bangladesh-are-still-without-homes-after-severe-floods.

19. National Centers for Environmental Information (NCEI), "Monthly Climate Reports | Wildfires Report | Annual," 2024, https://www.ncei.noaa.gov/access/monitoring/monthly-report/fire/202413.

20. World Economic Forum, "How Can We Bring 2.6 Billion People Online to Bridge the Digital Divide?," January 14, 2024, https://www.weforum.org/stories/2024/01/digital-divide-internet-access-online-fwa/.

21. Xiayin Zhang et al., "The Combination of Brain-Computer Interfaces and Artificial Intelligence: Applications and Challenges," *Annals of Translational Medicine* 8, no. 11 (June 2020): 712, https://doi.org/10.21037/atm.2019.11.109.

22. aiforgoodstg2, "Revolutionizing Mobility with Advanced AI-Powered Exoskeletons," AI for Good, June 24, 2024, https://aiforgood.itu.int/revolutionizing-mobility-with-advanced-ai-powered-exoskeletons/.

23. James Wright, "Inside Japan's Long Experiment in Automating Elder Care," *MIT Technology Review*, January 9, 2023, https://www.technologyreview.com/2023/01/09/1065135/japan-auto mating-eldercare-robots/.

24. National Association of Home Builders, *Is the Construction Workforce Older than Other Industries?* (blog), June 6, 2023, https://www.nahb.org/blog/2023/06/age-of-construction-workforce.

25. "Spot," Boston Dynamics, https://bostondynamics.com/products/spot/.

26. Derek Thompson, "The Anti-Social Century," *The Atlantic*, January 8, 2025, https://www.theatlantic.com/magazine/archive/2025/02/american-loneliness-personality-politics/681091/.

27. "Nearly 3 in 4 Teens Have Used AI Companions, New Survey Finds," Common Sense Media, https://www.commonsensemedia.org/press-releases/nearly-3-in-4-teens-have-used-ai-compan ions-new-national-survey-finds.

28. Shoshana Zuboff, *The Age of Surveillance Capitalism: The Fight for a Human Future at the New Frontier of Power* (PublicAffairs, 2020).

29. Neil Woodcock, "Prioritising AI over Climate Change Would Be Catastrophic," Tech Informed, October 28, 2024, https://techin formed.com/prioritising-ai-over-climate-change-would-be-catastrophic/.

30. Greg Epstein, "Silicon Valley's Obsession with AI Looks a Lot Like Religion," *The MIT Press Reader*, https://thereader.mit press.mit.edu/silicon-valleys-obsession-with-ai-looks-a-lot-like-religion/.

31. The American Psychoanalyst's Substack, "Computation Is Not Mentation," *The American Psychoanalyst's Substack* (blog), March 20, 2025, https://americanpsychoanalyst.substack.com/p/computation-is-not-mentation.

32. Gary Marcus, "Breaking News: Scaling Will Never Get Us to AGI," *Marcus on AI* (blog), April 8, 2024, https://garymar cus.substack.com/p/breaking-news-scaling-will-never.

33. "Zora," Memory Alpha, June 14, 2025, https://memory-alpha.fan dom.com/wiki/Zora.

34. Douglas Adams, *The Hitch Hiker's Guide to the Galaxy* (Pan Books, 1979).

35. Alokya Kanungo, "The Real Environmental Impact of AI," Earth. Org, July 18, 2023, https://earth.org/the-green-dilemma-can-ai-fulfil-its-potential-without-harming-the-environment/.

36. "Attention & Mental Health," Center for Humane Technology, https://www.humanetech.com/attention-mental-health.

37. Dana Goldstein, "American Children's Reading Skills Reach New Lows," *New York Times*, January 29, 2025, sec. U.S., https://www.nytimes.com/2025/01/29/us/reading-skills-naep.html.

38. James D. Walsh, "Everyone Is Cheating Their Way Through College," *New York Magazine*, May 7, 2025, https://nymag.com/intelligencer/article/openai-chatgpt-ai-cheating-education-college-students-school.html.

39. Nataliya Kosmyna et al., "Your Brain on ChatGPT: Accumulation of Cognitive Debt When Using an AI Assistant for Essay Writing Task," arXiv.org, June 10, 2025, https://arxiv.org/abs/2506.08872v1.

40. Nitish Pahwa, "Musk Built the MAGA Chatbot of His Dreams. Then It Wouldn't Shut Up About 'White Genocide'," *Slate*, May 16, 2025, https://slate.com/technology/2025/05/elon-musk-grok-chatbot-glitch-south-africa-white-genocide.html.

41. "Leonardo Da Vinci's Robotic Knight Invention," Leonardo Da Vinci Inventions, https://www.da-vinci-inventions.com/robotic-knight.

42. "Jaquet-Droz Automata," *Wikipedia*, January 24, 2025, https://en.wikipedia.org/w/index.php?title=Jaquet-Droz_automata&oldid=1271438021.

43. Sarah Baldwin, "Ada Lovelace and the Analytical Engine," *Ada Lovelace* (blog), July 26, 2018, https://blogs.bodleian.ox.ac.uk/adalovelace/2018/07/26/ada-lovelace-and-the-analytical-engine/.

44. A. M. Turing, "Computing Machinery and Intelligence," *Mind, New Series* 59, no. 236 (1950): 433–60, http://www.jstor.org/stable/2251299.

45. Dartmouth, "Artificial Intelligence (AI) Coined at Dartmouth," https://home.dartmouth.edu/about/artificial-intelligence-ai-coined-dartmouth.

46. "Switzerland Joins the AI Race with Open Multilingual LLM," Multilingual, July 15, 2025, https://multilingual.com/swiss-multi lingual-llm/.

47. "Article 3: Definitions," https://artificialintelligenceact.eu/article/3/.

48. Yordanka Ivanova, "Generative AI and the AI Act," https://www.europarl.europa.eu/cmsdata/277774/02.YordankaI vanova.pdf.

49. Neuralink, "Neuralink—Pioneering Brain Computer Interfaces," https://neuralink.com/.

50. Michael Cheng-Tek Tai, "The Impact of Artificial Intelligence on Human Society and Bioethics," *Tzu Chi Medical Journal* 32, no. 4 (2020): 339–43, https://doi.org/10.4103/tcmj.tcmj_71_20.

51. https://openai.com/our-structure/.

52. Matt O'Brien, "Tech Companies Want to Build Artificial General Intelligence. But Who Decides When AGI Is Attained?," AP News, April 4, 2024, https://apnews.com/article/agi-artificial-general-intelligence-existential-risk-meta-openai-deepmind-science-ff5662a056d3cf3c5889a73e929e5a34.

53. Sam Altman, "Reflections," *Sam Altman* (blog), January 5, 2025, https://blog.samaltman.com/reflections.

54. Sam Altman [@sama], "Twitter Hype Is out of Control Again," Twitter, January 20, 2025, https://x.com/sama/status/18812584 43669172470.

55. Nick Bostrom, "How Long Before SuperIntelligence," March 12, 2008, https://nickbostrom.com/superintelligence.

56. O'Brien, "Tech Companies Want to Build Artificial General Intelligence. But Who Decides When AGI Is Attained?"

57. OpenAI, "Planning for AGI and Beyond," February 24, 2023, https://openai.com/index/planning-for-agi-and-beyond/.

58. Daniel Tencer, "Ex-Google CEO Eric Schmidt: AI That Is 'as Smart as the Smartest Artist' Will Be Here in 3 to 5 Years," Music Business Worldwide, April 16, 2025, https://www.musicbusi nessworldwide.com/ex-google-ceo-eric-schmidt-ai-that-is-as-smart-as-the-smartest-artist-will-be-here-in-3-to-5-years/.

59. Ina Fried, "Meta Finalizes $15 Billion Deal for Scale AI Stake," Axios, June 12, 2025, https://www.axios.com/2025/06/13/meta-scale-ai-deal.

60. "AI Investment Trends 2025: Beyond the Bubble," J.P. Morgan Asset Management, November 19, 2024, https://am.jpmorgan .com/se/en/asset-management/per/insights/market-insights/ investment-outlook/ai-investment/.

61. "What Is AlphaFold?," EMBL-EBI, https://www.ebi.ac.uk/train ing/online/courses/alphafold/an-introductory-guide-to-its-strengths-and-limitations/what-is-alphafold/.

62. Brian Buntz, "7 Case Studies Highlighting the Potential of DeepMind's AlphaFold," Drug Discovery and Development, July 28, 2022, https://www.drugdiscoverytrends.com/7-ways-deep mind-alphafold-used-life-sciences/.

63. Social Media Victims Law Center, "Mom Sues AI Chatbot in Federal Lawsuit After Son's Death," October 25, 2024, https:// socialmediavictims.org/blog/lawsuit-filed-against-character-ai-after-teens-death/.

64. Cade Metz et al., "How Tech Giants Cut Corners to Harvest Data for AI," *New York Times*, April 6, 2024, https://www.nytimes .com/2024/04/06/technology/tech-giants-harvest-data-artificial-intelligence.html.

65. Chunsik Lee, Junga Kim, Joon Soo Lim, and Donghee Shin, "Generative AI Risks and Resilience: How Users Adapt to Hallucination and Privacy Challenges," *Telematics and Informatics Reports* 19 (September 2025), 100221, https://doi.org/10.1016/ j.teler.2025.100221.

66. "Lawyers Submitted Bogus Case Law Created by ChatGPT. A Judge Fined Them $5,000," AP News, June 22, 2023, https:// apnews.com/article/artificial-intelligence-chatgpt-fake-case-lawyers-d6ae9fa79d0542db9e1455397aef381c.

67. The King, on Application of Frederick Ayinde and the London Bourough of Haringey [2025] EWHC 1383 (admin), https://www .judiciary.uk/wp-content/uploads/2025/06/Ayinde-v-London-Borough-of-Haringey-and-Al-Haroun-v-Qatar-National-Bank.pdf.

68. Paul Mozur, "One Month, 500,000 Face Scans: How China Is Using A.I. to Profile a Minority," *New York Times*, April 14, 2019, https://www.nytimes.com/2019/04/14/technology/china-surveillance-artificial-intelligence-racial-profiling.html.

69. Daniel Zaleznic, "Facebook and Genocide: How Facebook Contributed to Genocide in Myanmar and Why It Will Not Be Held Accountable," Harvard Law School, Systemic Justice Project 1, 2021, https://systemicjustice.org/article/facebook-and-geno cide-how-facebook-contributed-to-genocide-in-myanmar-and-why-it-will-not-be-held-accountable/.

70. Kalley Huang, "Meta Curbs Privacy Teams' Sway over Product Releases," The Information, https://www.theinformation.com/articles/meta-curbs-privacy-teams-sway-over-product-releases.

71. "Project Gutenberg," Project Gutenberg, https://www.gutenberg.org/.

72. "Manu/Project_Gutenberg· Datasets at Hugging Face," October 9, 2024, https://huggingface.co/datasets/manu/project_gutenberg.

73. Martin Gerlach and Francesc Font-Clos, "A Standardized Project Gutenberg Corpus for Statistical Analysis of Natural Language and Quantitative Linguistics," *Entropy* 22, no. 1 (January 20, 2020): 126, https://doi.org/10.3390/e22010126.

74. Jane Austen, *Persuasion* (1817).

75. "Women in Translation," Brookline Booksmith, https://brookline booksmith.com/collections/women-translation.

76. "1977–1989 Timeline," Society for Women's Health Research, https://swhr.org/about/1977-1989-timeline/.

77. Amy Huebschmann and Judith Regensteiner, "Women Are at a Higher Risk of Dying from Heart Disease – in Part Because Doctors Don't Take Major Sex and Gender Differences into Account," *Journal of Health Economics and Outcomes Research*, November 12, 2024, https://jheor.org/post/2779-women-are-at-a-higher-risk-of-dying-from-heart-disease-in-part-because-doctors-don-t-take-major-sex-and-gender-differences-into-account.

78. Caroline Criado Perez, *Invisible Women* (Penguin, 2020), https://www.penguin.co.uk/books/435554/invisible-women-by-perez-caroline-criado/9781784706289.

79. Christian Mammen, "Why Human-Created Input Data Is Needed to Maintain AI Models," *The National Law Review* XV, no. 165 (December 3, 2024), https://natlawreview.com/article/why-human-created-input-data-needed-maintain-ai-models.

80. Gartner, "Gartner Business Insights, Strategies & Trends For Executives," https://www.gartner.com/en/insights.

81. "What Is a Digital Twin?," IBM Think, August 5, 2021, https://www.ibm.com/think/topics/what-is-a-digital-twin.

82. "The Social Dilemma - A Netflix Original Documentary," The Social Dilemma, https://thesocialdilemma.com/.

83. Emily Kwong and Rachel Carlson, "This Week in Science: A New Desert Flower, Virtual Lemonade and Prehistoric Bone Tools," Spokane Public Radio, March 6, 2025, https://www.spokanepublicradio.org/2025-03-06/this-week-in-science-a-new-desert-flower-virtual-lemonade-and-prehistoric-bone-tools.

84. *Surrogates* (Touchstone Pictures, Mandeville Films, Brownstone Productions (II), 2009).

85. *Ready Player One* (Warner Bros., Amblin Entertainment, Village Roadshow Pictures, 2018).

86. Todd B. Kashdan, "The Mental Benefits of Vacationing Somewhere New," *Harvard Business Review*, https://hbr.org/2018/01/the-mental-benefits-of-vacationing-somewhere-new.

87. Serious Games Society, "Welcome to the Serious Games Society," https://seriousgamessociety.org/.

88. Juliana Gómez-Quintero et al., "A Scoping Study of Crime Facilitated by the Metaverse," *Futures* 157 (March 1, 2024): 103338, https://doi.org/10.1016/j.futures.2024.103338.

89. Staff Writer, "Malicious Bots Drive Surge in Serious Human Crimes," *Arkose Labs* (blog), October 31, 2023, https://www.arkoselabs.com/blog/malicious-bots-drive-surge-in-serious-human-crimes/.

90. "How to Spot Bots on Social Media," Microsoft 365 Life Hacks, June 27, 2024, https://www.microsoft.com/en-us/microsoft-365-life-hacks/privacy-and-safety/how-to-spot-bots-on-social-media.

91. "Algorithm - Glossary," NIST Computer Security Resource Center, https://csrc.nist.gov/glossary/term/algorithm.

92. Brian Schatz, "Schatz, Cruz, Murphy, Britt Introduce Bipartisan Legislation to Keep Kids Safe, Healthy, off Social Media," January 28, 2025, https://www.schatz.senate.gov/news/press-releases/01/28/2025/schatz-cruz-murphy-britt-introduce-bipartisan-legislation-to-keep-kids-safe-healthy-off-social-media.

93. Emily Young, "Frequent Social Media Use and Experiences with Bullying Victimization, Persistent Feelings of Sadness or Hopelessness, and Suicide Risk Among High School Students – Youth Risk Behavior Survey, United States, 2023," *MMWR Supplements* 73 (2024), https://doi.org/10.15585/mmwr.su7304a3.

94. "The A.I. Dilemma," 2023, https://www.youtube.com/watch?v=xoVJKj8lcNQ.

95. Mark Hachman, "Microsoft Recall, Tested: I Like the AI, but I Just Can't Trust It," *PCWorld*, December 9, 2024, https://www.pcworld.com/article/2535272/microsoft-recall-windows-ai-tested-hands-on.html.

96. "6 Things to Learn from the Garmin Security Breach," *Terranova Security* (blog), February 3, 2023, https://www.terranovasecurity.com/blog/garmin-security-breach.

97. Lila MacLellan, "These Are the People Interested in Buying 23andMe," *Fortune*, April 1, 2025, https://fortune.com/2025/04/01/23andme-for-sale-interested-buyers-nucleus-sei-foundation/.

98. "Everything You Should Know About the Dark Web," *Tulane University School of Professional Advancement* (blog), https://sopa.tulane.edu/blog/everything-you-should-know-about-dark-web.

99. Aumyo Hassan and Sarah J. Barber, "The Effects of Repetition Frequency on the Illusory Truth Effect," *Cognitive Research: Principles and Implications* 6 (May 13, 2021): 38, https://doi.org/10.1186/s41235-021-00301-5.

100. Cade Metz, "Why Does OpenAI Need So Much Money?," *New York Times*, December 17, 2024, https://www.nytimes.com/2024/12/17/technology/openai-chatgpt-funding.html.

101. "Mexico FDI Disappoints Despite Hitting Record US$37bn," BNamericas.Com, February 27, 2025, https://www.bnamericas.com/en/analysis/mexico-fdi-disappoints-despite-hitting-record-us37bn.

102. Gary Marcus, "Deep Learning, Deep Scandal," *Marcus on AI* (blog), April 7, 2025, https://garymarcus.substack.com/p/deep-learning-deep-scandal.

103. Richard Fletcher and Rasmus Kleis Nielsen, *"What Does the Public in Six Countries Think of Generative AI in News?,"* Reuters Institute for the Study of Journalism, 2024, https://doi.org/10.60625/RISJ-4ZB8-CG87.

104. Julie Ray, "Americans Express Real Concerns About Artificial Intelligence," Gallup.Com, August 27, 2024, https://news.gallup.com/poll/648953/americans-express-real-concerns-artificial-intelligence.aspx.

105. Jamie Ballard, "Americans Are Increasingly Likely to Say AI Will Negatively Affect Society," YouGov, July 18, 2025, https://today.yougov.com/politics/articles/52615-americans-increasingly-likely-say-ai-artificial-intelligence-negatively-affect-society-poll.

106. "2025 Edelman Trust Barometer," Edelman, https://www.edelman.com/trust/2025/trust-barometer.

107. Jayme Quinn, "How Much of Communication Is Nonverbal?," UT Permian Basin Online, November 3, 2020, https://online.utpb.edu/about-us/articles/communication/how-much-of-communication-is-nonverbal.

108. Nancy Albinson, Sam Balaji, and Yang Chu, "Building Digital Trust: Technology Can Lead the Way," Deloitte Insights, n.d., https://www2.deloitte.com/content/dam/insights/us/articles/6320_Building-digital-trust/DI_Building-digital-trust.pdf.

Chapter 2

1. Tobias Mann, "US Senator Calls for Jail, Fines for Aiding Chinese AI," The Register, February 3, 2025, https://www.theregister.com/2025/02/03/us_senator_download_chinese_ai_model/.

2. Steve Stecklow et al., "Tesla Workers Shared Sensitive Images Recorded by Customer Cars," Reuters, April 6, 2023, https://www.reuters.com/technology/tesla-workers-shared-sensitive-images-recorded-by-customer-cars-2023-04-06/.

3. "Smart Toy Awards," https://www.smarttoyawards.org/.

4. Beth Maundrill, "Barbie's Data Privacy Scandal," *Infosecurity Magazine* (blog), July 27, 2023, https://www.infosecurity-magazine.com/blogs/barbies-data-privacy-scandal/.

5. Tim Medin, "Doll Hacking: The Good, The Bad(Words) and the Ugly (Features)," Red Siege Information Security, November 18,

2015, https://redsiege.com/tools-techniques/2015/11/doll-hacking-good-badwords-and-ugly/.

6. Finn Mystad, "Cayla forbudt i Tyskland, på salg i Norge," Forbrukerrådet, February 17, 2017, https://www.forbrukerra det.no/siste-nytt/cayla-forbudt-i-tyskland-pa-salg-i-norge/.

7. E. Beneteau, A. Boone, Y. Wu, J. A. Kientz, J. Yip, and A. Hiniker, "Parenting with Alexa: Exploring the Introduction of Smart Speakers on Family Dynamics," *Proceedings of the 2020 CHI conference on human factors in computing systems* (ACM, 2020), 1–13, https://doi.org/10.1145/3313831.3376344.

8. Justine Cassell, "Towards a Model of Technology and Literacy Development: Story Listening Systems," *Journal of Applied Developmental Psychology* 25, no. 1 (January–February 2004): 75–105, https://www.sciencedirect.com/science/article/abs/pii/S0193397303001369?via%3Dihub.

9. "FTC Grants Approval for New COPPA Verifiable Parental Consent Method," Federal Trade Commission, December 23, 2013, https://www.ftc.gov/news-events/news/press-releases/2013/12/ftc.grants-approval-new-coppa-verifiable-parental-con sent-method.

10. "What the FTC Does," Federal Trade Commission, September 11, 2018, https://www.ftc.gov/news-events/media-resources/what-ftc-does.

11. *United States v. Amazon.Com, Inc.*, 223-Cv-00811 - CourtListener .Com, CourtListener, https://www.courtlistener.com/docket/67455716/united-states-v-amazoncom-inc/.

12. "FTC and DOJ Charge Amazon with Violating Children's Privacy Law by Keeping Kids' Alexa Voice Recordings Forever and Undermining Parents' Deletion Requests," Federal Trade Commission, May 31, 2023, https://www.ftc.gov/news-events/news/press-releases/2023/05/ftc-doj-charge-amazon-violating-childrens-privacy-law-keeping-kids-alexa-voice-record ings-forever.

13. "Social Media and Youth Mental Health," The U.S. Surgeon General's Advisory, 2023, https://www.hhs.gov/sites/default/files/sg-youth-mental-health-social-media-advisory.pdf.

14. David Stillwell, "Why Facebook Knows You Better than Your Mum," Cambridge Judge Business School, December 15, 2015,

https://www.jbs.cam.ac.uk/2015/why-facebook-knows-you-better-than-your-mum/.

15. Stacey B. Steinberg, "The Myth of Children's Online Privacy Protection," *SMU Law Review* 77, no. 2 (2024): 441, https://doi.org/10.25172/smulr.77.2.8.

16. Jeff Horwitz and Aaron Tilley, "Apple Helped Nix Part of a Child Safety Bill. More Fights Are Expected," *Wall Street Journal*, September 2, 2024, https://www.wsj.com/tech/apple-helped-nix-part-of-a-child-safety-bill-more-fights-are-expected-23905d4d.

17. "Meta Prepares Boarder Roll-out of Age Verifications Tool to Meet New Compliance Standards" Digital Information World WebDesk 7/17/2025, https://www.digitalinformationworld.com/2025/07/meta-prepares-broader-rollout-of-age.html#:~:text=Broader%20Impact%20on%20Platforms,%2C%20illness%2C%20or%20personal%20beliefs.

18. Ted Cruz, S.146 - 119th Congress (2025–2026): TAKE IT DOWN Act," 2025, https://www.congress.gov/bill/119th-congress/senate-bill/146.

19. Online Safety Act 2023, https://www.legislation.gov.uk/ukpga/2023/50.

20. "Molly Russell: Prevention of Future Deaths Report," UK Courts and Tribunals Judiciary, October 13, 2022, https://www.judiciary.uk/prevention-of-future-death-reports/molly-russell-prevention-of-future-deaths-report/.

21. Department of Justice, Government of Canada, Bill C-63: An Act to Enact the Online Harms Act, to Amend the Criminal Code, the Canadian Human Rights Act and An Act Respecting the Mandatory Reporting of Internet Child Pornography by Persons Who Provide an Internet Service and to Make Consequential and Related Amendments to Other Acts, May 30, 2024, https://www.justice.gc.ca/eng/csj-sjc/pl/charter-charte/c63.html.

22. "How Big Is Your Vocabulary?," Toppan Digital Language, https://toppandigital.com/translation-blog/how-big-is-your-vocabulary/.

23. "When Do Babies Start Talking? – Children's Health," https://www.childrens.com/health-wellness/when-do-babies-start-talking.

24. "New Research Sheds Light on the US Shift Toward Smaller Families," The Institute for Social Research Population Studies Center, October 14, 2024, https://psc.isr.umich.edu/news/new-research-sheds-light-on-the-us-shift-toward-smaller-families/.

25. https://www.unicef.org/child-rights-convention/convention-text.

26. "Policy Guidance on AI for Children - Draft 1.0," UNICEF, September 2020, https://www.unicef.org/innocenti/media/1326/file/UNICEF-Global-Insight-policy-guidance-AI-children-draft-1.0-2020.pdf.

27. Mike Wall, "SpaceX Will Start Launching Starships to Mars in 2026, Elon Musk Says," Space.com, September 8, 2024, https://www.space.com/spacex-starship-mars-launches-2026-elon-musk.

28. *Elysium* (TriStar Pictures, Media Rights Capital (MRC), QED International, 2013).

29. *Don't Look Up* (Hyperobject Industries, 2021).

Chapter 3

1. Ed Bruce, "Mammas Don't Let Your Babies Grow up to Be Cowboys," United Artists, 1975.

2. Michael Greenstone et al., "Thirteen Economic Facts About Social Mobility and the Role of Education," The Brookings Institution, June 26, 2013, https://www.brookings.edu/articles/thirteen-economic-facts-about-social-mobility-and-the-role-of-education/.

3. Jennifer Gunn, "History and Evolution of STEAM Learning in the United States," Resilient Educator, November 3, 2017, https://resilienteducator.com/classroom-resources/evolution-of-stem-and-steam-in-the-united-states/.

4. Pascal Bornet, "The Hidden Biases of AI: An Eye-Opening Experiment in a Divided World," *IRREPLACEABLE with AI* (blog), February 13, 2025, https://pascalbornet.substack.com/p/the-hidden-biases-of-ai-an-eye-opening.

5. Tor Constantino, "Women Make Up 29% of the AI Workforce — Here's How to Fix It," *Forbes*, November 12, 2024, https://www.forbes.com/sites/torconstantino/2024/11/12/women-make-up-29-of-the-ai-workforce---heres-how-to-fix-it/.

6. https://www.aaha.org/newstat/publications/charts-the-state-of-women-in-vet-med/.

7. "The 2024 Veterinary Industry Stats You Need to Know," June 10, 2024, https://petdesk.com/blog/2024-veterinary-industry-stats/.

8. "Share of Female Physicians by State U.S. 2024," Statista, https://www.statista.com/statistics/1100676/share-of-female-physicians-across-us-states/.

9. Ali Shehab, "Is AI Ruining Your Kind's Critical Thinking?, *Psychology Today*, https://www.psychologytoday.com/us/blog/the-human-algorithm/202504/is-ai-ruining-your-kids-critical-thinking.

10. Lennart Meincke, Gideon Nave, and Christian Terwiesch, "ChatGPT Decreases Idea Diversity in Brainstorming," *Nature Human Behaviour* (May 14, 2025): 1–3, https://doi.org/10.1038/s41562-025-02173-x.

11. "Education and Economic Growth," Theirworld, https://key.theirworld.org/resources/economic-growth.

12. Jeanne Carroll, "Are Calculators Killing Our Ability to Work It Out in Our Head?," ABC News, August 9, 2015, https://www.abc.net.au/news/2015-08-10/are-calculators-affecting-maths-skills/6684542.

13. Steven Yoder, "Math Ends the Education Careers of Thousands of Community College Students. A Few Schools Are Trying Something New," June 24, 2024, https://hechingerreport.org/math-ends-the-education-careers-of-thousands-of-community-college-students-a-few-schools-are-trying-something-new/.

14. "TIMSS," Institution of Educational Sciences, https://nces.ed.gov/timss/.

15 Grace Chen, "Why Do 60% of Community College Students Need Remedial Coursework?," *Community College Review* (blog), March 21, 2022, https://www.communitycollegereview.com/blog/why-do-60-of-community-college-students-need-remedial-coursework.

16. Ying Xu et al., "Growing Up with Artificial Intelligence: Implications for Child Development," in *Handbook of Children and Screens: Digital Media, Development, and Well-Being from Birth Through Adolescence*, ed. Dimitri A. Christakis and Lauren Hale (Springer Nature Switzerland, 2025), 611–17, https://doi.org/10.1007/978-3-031-69362-5_83.

17. Gabriela Galvin, "AI Models Think All Nurses Are Women and Senior Doctors Are Men," Euronews, September 23, 2024, https://www.euronews.com/health/2024/09/23/ai-models-like-chatgpt-think-nurses-are-women-and-senior-doctors-are-men-study-shows.

18. Meincke, Nave, and Terwiesch, "ChatGPT Decreases Idea Diversity in Brainstorming"; N. Kosmyna et al., "Your Brain on ChatGPT."

19. Roberta Michnick Golinkoff et al., "Language Matters: Denying the Existence of the 30-Million-Word Gap Has Serious Consequences," *Child Development* 90, no. 3 (2019): 985–92, https://doi.org/10.1111/cdev.13128.

20. Ying Xu et al., "Dialogue with a Conversational Agent Promotes Children's Story Comprehension via Enhancing Engagement," *Child Development* 93, no. 2 (March 2022), https://doi.org/10.1111/cdev.13708.

21. "10 Ways Play Impacts Child Development," Kid Care Pediatrics, July 23, 2024, https://kidcarepediatrics.com/the-importance-of-play-for-children/.

22. James D. Walsh, "Everyone Is Cheating Their Way Through College," Intelligencer, May 7, 2025, https://nymag.com/intelligencer/article/openai-chatgpt-ai-cheating-education-college-students-school.html.

23. David Brooks, "We Are the Most Rejected Generation," *New York Times*, May 15, 2025, https://www.nytimes.com/2025/05/15/opinion/rejection-college-youth.html.

24. Sherin Shibu, "AI Startup Anthropic to Job Seekers: No AI on Applications," *Entrepreneur*, February 5, 2025, https://www.entrepreneur.com/business-news/ai-startup-anthropic-to-job-seekers-no-ai-on-applications/486645.

25. Eva Chan and Geoffrey Scott, "2024 Hiring Trends Survey: What Makes a Great Job Candidate?," *Resume Genius* (blog), June 13, 2024, https://resumegenius.com/blog/job-hunting/hiring-trends-survey.

26. "Can Playing Piano Help Prevent Alzheimer's Disease?," *Riverside Music Studios* (blog), October 13, 2023, https://riversidemusicstudios.com/2023/10/can-playing-piano-help-prevent-alzheimers-disease/.

27. "Study Shows Learning a Second Language Thwarts Onset of Dementia," *LAS News*, Iowa State University, January 28, 2021,

https://news.las.iastate.edu/2021/01/28/study-shows-learning-a-second-language-thwarts-onset-of-dementia/.

28. Jeremias Ivan, Rizky Nurdiansyah, and Arli Aditya Parikesit, "Mathematical Problem Solving: One Way to Prevent Dementia," *Frontiers in Health Informatics* 8, no. 1 (May 13, 2019): 10, https://doi.org/10.30699/fhi.v8i1.179.

29. "Final Draft of EU AI Act Leaked," Hunton, February 1, 2024, https://www.hunton.com/privacy-and-information-security-law/final-draft-of-eu-ai-act-leaked.

30. Léo Boisvert et al., "WorkArena++: Towards Compositional Planning and Reasoning-Based Common Knowledge Work Tasks," arXiv, 2024, https://doi.org/10.48550/ARXIV.2407.05291.

31. Dustin Chambers, "Their Water Taps Ran Dry When Meta Built Next Door," *New York Times*, July 16, 2025, https://www.nytimes.com/2025/07/14/technology/meta-data-center-water.html.

32. Renee Cho, "AI's Growing Carbon Footprint – State of the Planet," Columbia Climate School, June 9, 2023, https://news.climate.columbia.edu/2023/06/09/ais-growing-carbon-footprint/.

33. Adam Zewe, "Explained: Generative AI's Environmental Impact," *MIT News*, January 17, 2025, https://news.mit.edu/2025/explained-generative-ai-environmental-impact-0117.

34. Arman Shehabi et al., "*2024 United States Data Center Energy Usage Report*," Lawrence Berkeley National Laboratory, December 2024, https://eta-publications.lbl.gov/sites/default/files/2024-12/lbnl-2024-united-states-data-center-energy-usage-report_1.pdf.

35. Chase DiBenedetto, "Google's Former CEO: Companies Shouldn't Let Climate Concerns Slow AI Advances," Mashable, October 7, 2024, https://mashable.com/article/former-google-ceo-invest-ai-despite-climate-concerns.

36. Joe Wilkins, "Former Google CEO Tells Congress That 99 Percent of All Electricity Will Be Used to Power Superintelligent AI," Yahoo News, April 12, 2025, https://www.yahoo.com/news/former-google-ceo-tells-congress-164530047.html.

37. Gavin Schmidt and Zeke Hausfather, "We Study Climate Change. We Can't Explain What We're Seeing," *New York Times*, November 13, 2024, https://www.nytimes.com/2024/11/13/opinion/climate-change-heat-planet.html.

38. United Nations Environment Programme et al., "Emissions Gap Report 2024: No More Hot Air . . . Please! With a Massive Gap between Rhetoric and Reality," Countries Draft New Climate Commitments, United Nations Environment Programme, 2024, https://doi.org/10.59117/20.500.11822/46404.

39. Ezra Klein, "Now Is the Time of Monsters," *New York Times*, January 12, 2025, https://www.nytimes.com/2025/01/12/opinion/ai-climate-change-low-birth-rates.html.

Chapter 4

1. Kevin Roose, "Can A.I. Be Blamed for a Teen's Suicide?," *New York Times*, October 23, 2024, https://www.nytimes.com/2024/10/23/technology/characterai-lawsuit-teen-suicide.html.

2. Julia Carrie Wong, "Loathe Thy Neighbor: Elon Musk and the Christian Right Are Waging War on Empathy," *The Guardian*, April 8, 2025, https://www.theguardian.com/us-news/ng-interactive/2025/apr/08/empathy-sin-christian-right-musk-trump.

3. Catrin Misselhorn, "'Empathetic' AI Has More to Do with Psychopathy Than Emotional Intelligence – but That Doesn't Mean We Can Treat Machines Cruelly," The Conversation, March 21, 2024, http://theconversation.com/empathetic-ai-has-more-to-do-with-psychopathy-than-emotional-intelligence-but-that-doesnt-mean-we-can-treat-machines-cruelly-225216.

4. "10 Questions in 10 Minutes: Ex Machina Film Advisor, Professor Murray Shanahan," *The Human Podcast*, https://rss.com/podcasts/thehumanpodcastrss/1865873/.

5. Sarah Perez, "Tinder Will Try AI-Powered Matching as the Dating App Continues to Lose Users," TechCrunch, February 6, 2025, https://techcrunch.com/2025/02/06/tinder-will-try-ai-powered-matching-as-the-dating-app-continues-to-lose-users/.

6. Magdalene J. Taylor, "It's Not You: Dating Apps Are Getting Worse," *New York Times*, March 16, 2024, https://www.nytimes.com/2024/03/16/opinion/dating-apps-hinge-tinder-bumble.html.

7. Camila Barbagallo and Rocio Gonzalez Lantero, "Dating Apps' Darkest Secret: Their Algorithm," *IE HST Rewire Magazine*, https://rewire.ie.edu/dating-apps-darkest-secret-algorithm/.

8. Emily A. Vogels and Colleen McClain, "Key Findings about Online Dating in the U.S.," Pew Research Center, February 2, 2023, https://www.pewresearch.org/short-reads/2023/02/02/key-findings-about-online-dating-in-the-u-s/.

9. "Why Investors Might Still Swipe Right for Dating Apps," Morgan Stanley, April 20, 2023, https://www.morganstanley.com/ideas/online-dating-2030.

10. Cory Doctorow, "The 'Enshittification' of TikTok," *Wired*, January 23, 2023, https://www.wired.com/story/tiktok-platforms-cory-doctorow/.

11. "Lemon (Automobile)," *Wikipedia*, January 2, 2025, https://en.wikipedia.org/w/index.php?title=Lemon_(automobile)&oldid=1266818488.

12. Derek Thompson, "The Anti-Social Century," *The Atlantic*, January 8, 2025.

13. Anna Mae Duane, "AI 'Companions' Promise to Combat Loneliness, but History Shows the Danger of One-Way Relationships," UConn Today, February 20, 2024, https://today.uconn.edu/2024/02/ai-companions-promise-to-combat-loneliness-but-history-shows-the-danger-of-one-way-relationships/.

14. "Our Epidemic of Loneliness and Isolation," U.S. Surgeon General's Advisory on the Healing Effects of Social Connection and Community, 2023, https://www.hhs.gov/sites/default/files/surgeon-general-social-connection-advisory.pdf.

15. "Marriage and Men's Health," Harvard Health, June 5, 2019, https://www.health.harvard.edu/mens-health/marriage-and-mens-health.

16. Gale Berkowitz, "UCLA Study on Friendship Among Women," Women's Brain Health Initiative, January 2025, https://womensbrainhealth.org/think-tank/think-twice/ucla-study-on-friendship-among-women.

17. W. L. S. Burrough, "Evaluation of Feedback and a Graphical Game Element in an SMS Intervention to Increase Attendance Among At-Risk High School Students," Coventry University, 2018, https://pureportal.coventry.ac.uk/en/studentTheses/evaluation-of-feedback-and-a-graphical-game-element-in-an-sms-int.

18. Yuval Noah Harari, *21 Lessons for the 21st Century* (Random House, 2018).

19. "Eugenics and Scientific Racism," National Human Genome Research Institute, https://www.genome.gov/about-genomics/fact-sheets/Eugenics-and-Scientific-Racism.

20. Henry Ajder et al., *"The State of Deepfakes: Landscape, Threats, and Impact,"* Deep Trace Labs, September 2019, https://reg media.co.uk/2019/10/08/deepfake_report.pdf.

21. Elizabeth Laird, Maddy Dwyer, and Kristen Woelfel, *"In Deep Trouble - Surfacing Tech-Powered Sexual Harassment in K–12 Schools,"* Center for Democracy and Technology, September 2024, https://cdt.org/wp-content/uploads/2024/09/2024-09-26-final-Civic-Tech-Fall-Polling-research-1.pdf.

22. Linda Robinson, "Excerpts from a Conversation with Julie Inman Grant, Australia's eSafety Commissioner," *Council on Foreign Relations* (blog), December 20, 2024, https://www.cfr.org/blog/excerpts-conversation-julie-inman-grant-australias-esafety-commissioner.

23. K. C. Halm, John D. Seiver, and Edlira Kuka, "Who's Liable for Deepfakes? FTC Proposes to Target Developers of Generative AI Tools in Addition to Fraudsters," *Davis Wright Tremaine LLP* (blog), February 22, 2024, https://www.dwt.com/blogs/artificial-intelligence-law-advisor/2024/02/ftc-targets-tech-companies-for-generative-ai-fraud.

Chapter 5

1. Stephen Hawking, Stuart Russell, Max Tegmark, and Frank Wilczek, "Transcending Complacency on Superintelligent Machines," *Huffington Post*, April 19, 2014.

2. *The Theory of Everything* (Working Title Films, Dentsu Motion Pictures, Fuji Television Network (Fuji TV), 2014).

3. Peter Norvig and Stuart Russell, *Artificial Intelligence: A Modern Approach* (Pearson, 2022).

4. "The Reith Lectures: Stuart Russell - Living with Artificial Intelligence: AI: A Future for Humans," BBC Sounds, https://www.bbc.co.uk/sounds/play/m0012q21.

5. Stuart Russell, *Human Compatible: Artificial Intelligence and the Problem of Control* (Viking, 2019).

6. "Asilomar AI Principles," Future of Life Institute, August 11, 2017, https://futureoflife.org/open-letter/ai-principles/.

7. Max Tegmark, *Life 3.0: Being Human in the Age of Artificial Intelligence* (Vintage, 2018).

8. "Convention on Certain Conventional Weapons - Group of Governmental Experts on Lethal Autonomous Weapons Systems (2023)," United Nations, https://meetings.unoda.org/ccw-/convention-on-certain-conventional-weapons-group-of-governmental-experts-on-lethal-autonomous-weapons-systems-2023.

9. *Slaughterbots*, 2019, https://www.youtube.com/watch?v=O-2tpwW0kmU.

10. "Killer Robots: UN Vote Should Spur Action on Treaty," *Human Rights Watch*, January 3, 2024, https://www.hrw.org/news/2024/01/03/killer-robots-un-vote-should-spur-action-treaty.

11. "Milestones in the History of U.S. Foreign Relations," Office of the Historian, https://history.state.gov/milestones/1961-1968/npt.

12. Bruce Anderson, "Bruce Anderson: Auschwitz Is No Subject for Point-Scoring, and Neither," *The Independent*, February 25, 2008, https://www.the-independent.com/voices/commentators/bruce-anderson/bruce-anderson-auschwitz-is-no-subject-for-pointscoring-and-neither-party-emerges-with-any-credit-786861.html.

13. Samuel Bendett and David Kirichenko, "Battlefield Drones and the Accelerating Autonomous Arms Race in Ukraine," Modern War Institute at West Point, January 10, 2025, https://mwi.westpoint.edu/battlefield-drones-and-the-accelerating-autonomous-arms-race-in-ukraine/.

14. "Hummingbird Nano Air Vehicle (NAV)," Airforce Technology, https://www.airforce-technology.com/projects/hummingbird-nano-air-vehicle/.

15. Marc Santora et al., "A Thousand Snipers in the Sky: The New War in Ukraine," *New York Times*, March 3, 2025, https://www.nytimes.com/interactive/2025/03/03/world/europe/ukraine-russia-war-drones-deaths.html.

16. "Drones Now Rule the Battlefield in the Ukraine-Russia War," *New York Times*, https://www.nytimes.com/interactive/2025/03/03/world/europe/ukraine-russia-war-drones-deaths.html.

17. Patrick Tucker, "New AI-Powered Strike Drone Shows How Quickly Battlefield Autonomy Is Evolving," Defense One, October 10, 2024, https://www.defenseone.com/technology/2024/10/new-ai-powered-strike-drone-shows-how-quickly-battlefield-autonomy-evolving/400179/.

18. "Ukraine Strikes Russian Targets with Locally-Made Dovbush T10 Multirole Carrier Drone," Army Recognition Group, November 20, 2024, https://armyrecognition.com/focus-analysis-conflicts/army/conflicts-in-the-world/russia-ukraine-war-2022/ukraine-designs-locally-made-dovbush-t10-uav-able-to-carry-6-fpv-strike-drones.

19. Lily Jamali, "On Patrol at Mar-a-Lago, Robotic Dogs Have Their Moment," *BBC News*, November 17, 2024, https://www.bbc.com/news/articles/c30p16gn3pvo.

20. "A Diplomat's Guide to Autonomous Weapons Systems," Future of Life Institute, August 4, 2024, https://futureoflife.org/wp-content/uploads/2024/08/AWS-Guide-for-Diplomats_17-September-2024.pdf.

21. Tom Simonite, "When It Comes to Gorillas, Google Photos Remains Blind," Wired, January 11, 2018, https://www.wired.com/story/when-it-comes-to-gorillas-google-photos-remains-blind/.

22. Marissa Gerchick and Matt Cagle, "When It Comes to Facial Recognition, There Is No Such Thing as a Magic Number," *American Civil Liberties Union* (blog), February 7, 2024, https://www.aclu.org/news/privacy-technology/when-it-comes-to-facial-recognition-there-is-no-such-thing-as-a-magic-number.

23. David Jeans, "CEO Of Facial Recognition Company Clearview AI Resigns," *Forbes*, February 19, 2025, https://www.forbes.com/sites/davidjeans/2025/02/19/clearview-ai-ceo-resigns/.

24. "Biometric Data Becomes New Weapon in Hong Kong Protests," PBS News, July 27, 2019, https://www.pbs.org/newshour/

show/biometric-data-becomes-new-weapon-in-hong-kong-protests.

25. https://www.vatican.va/roman_curia/congregations/cfaith/documents/rc_ddf_doc_20250128_antiqua-et-nova_en.html.

Chapter 6

1. Dylan Walsh, "A New Look at the Economics of AI," MIT Management Sloan School, January 21, 2025, https://mit sloan.mit.edu/ideas-made-to-matter/a-new-look-economics-ai.

2. Lara Braff and Katie Nelson, "Chapter 15: The Global North: Introducing the Region," *Gendered Lives* (Milnes Open Textbooks, 2021), https://milnepublishing.geneseo.edu/genderedlives/chapter/chapter-15-the-global-north-introducing-the-region/.

3. Brandon Wegrowski, "Deforestation in the Amazon Rainforest," Ballard Brief, Fall 2019, https://ballardbrief.byu.edu/issue-briefs/deforestation-in-the-amazon-rainforest.

4. Dylan Walsh, "A New Look at the Economics of AI."

5. Michael Osborn and Carl Benedikt Frey, "Automation and the Future of Work – Understanding the Numbers," *Oxford Martin School* (blog), April 13, 2018, https://www.oxfordmartin.ox.ac.uk/blog/automation-and-the-future-of-work-understanding-the-numbers.

6. Evan Andrews, "Who Were the Luddites?," History, August 7, 2015, https://www.history.com/articles/who-were-the-luddites.

7. United Nations Educational, Scientific and Cultural Organization, OECD, and Inter-American Development Bank, *The Effects of AI on the Working Lives of Women* (OECD, 2022), https://doi.org/10.1787/14e9b92c-en.

8. Jim VandeHei and Mike Allen, "Behind the Curtain: Top AI CEO Foresees White-Collar Bloodbath," Axios, May 28, 2025, https://www.axios.com/2025/05/28/ai-jobs-white-collar-unemployment-anthropic.

9. Cody Corrall, Alyssa Stringer, and Kate Park, "A Comprehensive List of 2025 Tech Layoffs," TechCrunch, June 17, 2025, https://techcrunch.com/2025/06/17/tech-layoffs-2025-list/.

10. Swapna Venugopal Ramaswamy, "Barack Obama and Steve Bannon Agree on Something: AI's Role in American Jobs,

Politics," *USA Today*, June 2, 2025, https://www.usatoday.com/ story/news/politics/2025/06/02/obama-bannon-ai-white-collar-unemployment/83991339007/.

11. "The Influenza Epidemic of 1918," National Archives and Records Administration, https://www.archives.gov/exhibits/influenza-epidemic/.

12. Drew Gilpin Faust, *This Republic of Suffering by Drew Gilpin Faust* (Penguin Random House, 2009), https://www.penguin randomhouse.com/books/48460/this-republic-of-suffering-by-drew-gilpin-faust/.

13. Jonathan Reisman, "I'm a Doctor. ChatGPT's Bedside Manner Is Better Than Mine," *New York Times*, October 5, 2024, https://www.nytimes.com/2024/10/05/opinion/ai-chatgpt-medicine-doctor.html.

14. Kiran Tomlinson, Sonia Jaffe, Will Wang, Scott Counts, and Siddharth Suri, "Working with AI: Measuring the Occupational Implications of Generative AI," July 22, 2025, https://arxiv.org/abs/2507.07935.

15. Eli Tan, "At the Intersection of A.I. and Spirituality," *New York Times*, January 3, 2025, https://www.nytimes.com/2025/01/03/technology/ai-religious-leaders.html.

16. "Antiqua et Nova: Note on the Relationship Between Artificial Intelligence and Human Intelligence," Vatican, January 28, 2025, https://www.vatican.va/roman_curia/congregations/cfaith/documents/rc_ddf_doc_20250128_antiqua-et-nova_en.html.

17. Richard D. Land, "Pope Leo XIV, AI and Who and What Is a Human Being?," May 16, 2025, https://www.christianpost.com/voices/pope-leo-xiv-ai-and-who-and-what-is-a-human-being.html.

18. Rebecca Bellan, "Pope Leo Makes AI's Threat to Humanity a Signature Issue," TechCrunch, June 18, 2025, https://techcrunch.com/2025/06/18/pope-leo-makes-ais-threat-to-humanity-a-signature-issue/.

19. https://aiandfaith.org/events/ai-the-church-summit-three-mainline-denominations-start-the-conversation-together/.

20. Ana Khoirunisa et al., "Islam in the Midst of AI (Artificial Intelligence) Struggles: Between Opportunities and Threats,"

SUHUF 35, no. 1 (May 30, 2023): 45–52, https://doi.org/10.23917/suhuf.v35i1.22365.

21. Dan Friedman, "As AI Charges Ahead, Jewish Thinkers Are Falling Behind," *The Jewish News of Northern California*, March 21, 2025, https://jweekly.com/2025/03/21/as-ai-charges-ahead-jewish-thinkers-are-falling-behind/.

22. Chinmay Pandya, "A Hindu Perspective on AI Risks and Opportunities," Future of Life Institute, May 20, 2024, https://futureoflife.org/religion/a-hindu-perspective-on-ai-risks-and-opportunities/.

23. Peter D. Hershock, "A Buddhist Perspective on AI: Cultivating Freedom of Attention and True Diversity in an AI Future," Future of Life Institute, January 20, 2025, https://futureoflife.org/religion/a-buddhist-perspective-on-ai/.

24. Michael Marks, "A Legal Filing by the Houston Housing Authority Is Full of Fake Quotations," *Texas Standard*, https://www.texasstandard.org/stories/houston-housing-authority-fake-quotes-legal-brief/.

25. https://www.bailii.org/ew/cases/EWHC/Admin/2025/1383.html.

26. Matthew Lee, "Negligent Use or Negligent Refusal? Insights from Two Judicial Speeches on AI's Future by the Master of the Rolls, Sir Geoffrey Vos.," Natural and Artificial Intelligence in Law, https://naturalandartificiallaw.com/negligent-use-or-negligent-refusal-insights-from-two-judicial-speeches-on-ais-future-by-the-master-of-the-rolls-sir-geoffrey-vos/.

27. Amitai Etzioni and Oren Etzioni, "Keeping AI Legal," *Vanderbilt Journal of Entertainment and Technology Law* 19, no. 1 (2020): 133, https://scholarship.law.vanderbilt.edu/jetlaw.vol19/iss1/5.

28. Attributed to Lucas Cranach the Elder, "The Judgement of Solomon," 1519, https://www.rct.uk/collection/406032/the-judgement-of-solomon.

29. Kyle Wiggers, "OpenAI Is Funding Research into 'AI Morality'," *TechCrunch*, November 22, 2024, https://techcrunch.com/2024/11/22/openai-is-funding-research-into-ai-morality/.

30. Harrison Brooks, "Morgan Stanley's Job Cuts: A Strategic Move or a Sign of Troubled Times?," AInvest, March 21, 2025, https://www.ainvest.com/news/morgan-stanley-job-cuts-strategic-move-sign-troubled-times-2503/.

31. Chris Westfall, "How AI Revolution Is Driving 200,000 Layoffs On Wall Street," *Forbes*, January 13, 2025, https://www.forbes.com/sites/chriswestfall/2025/01/13/how-ai-revolution-is-driving-200000-layoffs-on-wall-street/.

32. Emmalyse Brownstein, "AI Could Automate More Than Half of Banking Jobs, Say CITI," *Business Insider*, June 19, 2024, https://www.businessinsider.com/ai-could-automate-more-than-half-of-banking-jobs-citi-2024-6.

33. David Brooks, "We Are the Most Rejected Generation," *New York Times*, May 15, 2025, https://www.nytimes.com/2025/05/15/opinion/rejection-college-youth.html.

34. Ibid.

35. Lindsay Ellis, "'You're Fighting AI with AI': Bots Are Breaking the Hiring Process," *Wall Street Journal*, May 11, 2024, https://www.wsj.com/lifestyle/careers/ai-job-application-685f29f7.

36. Alaina Demopoulos, "The Job Applicants Shut Out by AI: 'The Interviewer Sounded Like Siri'," *The Guardian*, March 6, 2024, sec. Technology, https://www.theguardian.com/technology/2024/mar/06/ai-interviews-job-applications.

37. Zahid Hussain et al., "Synergistic Integration of AI and 6-DOF Robotic Arm for Enhanced Bomb Disposal Capabilities," *Journal of The Institution of Engineers (India): Series C* 105, no. 5 (October 2024): 1247–61, https://doi.org/10.1007/s40032-024-01114-3.

38. PrivSec Report, "PwC Develops Facial Recognition Tools to Monitor Employees," GRC World Forums, June 18, 2020, https://www.grcworldforums.com/privacy-and-technology/pwc-develops-facial-recognition-tools-to-monitor-employees/54.article.

39. "Hewlett Packard Enterprise," https://www.hpe.com/us/en/home.html.

40. Walk Free, https://www.walkfree.org/.

41. "New Frontiers: The Use of Generative Artificial Intelligence to Facilitate Trafficking in Persons," Organization for Security and Co-operation in Europe, 2024, https://www.osce.org/cthb/579715.

42. Erin Kilbride, "Using AI to Fight Trafficking Is Dangerous," Human Rights Watch, July 1, 2024, https://www.hrw.org/news/2024/07/01/using-ai-fight-trafficking-dangerous.

43. "Human Trafficking and Social Media: A Modern Tool for Exploitation," HOPE, March 22, 2025, https://www.hopeagainst trafficking.org/human-trafficking-and-social-media-a-modern-tool-for-exploitation.

44. Kay Firth-Butterfield, *Human Trafficking and Human Rights* (Kendall Hunt Publishing, 2012).

45. "Gabby Salazar I Photographer and Author," https://www.gab bysalazar.com/about.

46. "Refik Anadol Studio," Refik Anadol Studio, https://refikanadol studio.com/.

47. "'A Machine-Shaped Hand': Read a Story from OpenAI's New Creative Writing Model," *The Guardian*, March 12, 2025, sec. Books, https://www.theguardian.com/books/2025/mar/12/a-machine-shaped-hand-read-a-story-from-openais-new-crea tive-writing-model.

48. Julia Sturges, "Taylor Swift, Deepfakes, and the First Amendment: Changing the Legal Landscape for Victims of Non-Consensual Artificial Pornography," *The Georgetown Journal of Gender and the Law* XXV, no. 2 (2024), https://www.law.georgetown.edu/gender-journal/online/volume-xxv-online/taylor-swift-deep fakes-and-the-first-amendment-changing-the-legal-landscape-for-victims-of-non-consensual-artificial-pornography/.

49. Dan Milmo and Alex Hern, "From Pope's Jacket to Napalm Recipes: How Worrying Is AI's Rapid Growth?," *The Guardian*, April 23, 2023, https://www.theguardian.com/technology/2023/apr/23/pope-jacket-napalm-recipes-how-worrying-is-ai-rapid-growth.

50. Martina Jaureguy, "Mauricio Macri Denounces Deepfake AI Videos Spread by Pro-Milei Trolls on Election Day," *Buenos Aires Herald*, May 18, 2025, https://buenosairesherald.com/politics/mauricio-macri-denounces-deepfake-ai-videos-spread-by-pro-milei-trolls-on-election-day.

51. https://www.cnn.com/2024/02/04/asia/deepfake-cfo-scam-hong-kong-intl-hnk/index.html.

52. Yuan Zhao et al., "Proactive Deepfake Defence via Identity Watermarking," in 2023 *IEEE/CVF Winter Conference on Applications of Computer Vision (WACV) (2023 IEEE/CVF Winter Conference on Applications of Computer Vision (WACV))*,

Waikoloa, HI, 2023), 4591–600, https://doi.org/10.1109/WACV56688.2023.00458.

53. Lovro Persen, "Fraud Detection: How Anti-Spoofing Facial Recognition Works," *IDnow* (blog), https://www.idnow.io/blog/fraud-detection-anti-spoofing-facial-recognition/.

54. Smita Khairnar et al., "Face Liveness Detection Using Artificial Intelligence Techniques: A Systematic Literature Review and Future Directions," *Big Data and Cognitive Computing* 7, no. 1 (March 2023): 37, https://doi.org/10.3390/bdcc7010.37.

55. "Deep Learning Architectures," IBM Developer, April 25, 2024, https://developer.ibm.com/articles/cc-machine-learning-deep-learning-architectures/.

56. "Governor Newsom Signs Bills to Combat Deepfake Election Content," *Governor of California* (blog), September 17, 2024, https://www.gov.ca.gov/2024/09/17/governor-newsom-signs-bills-to-combat-deepfake-election-content/.

Chapter 7

1. "Tattoo" (Episode), *Memory Alpha*, June 14, 2025, https://memory-alpha.fandom.com/wiki/Tattoo_(episode).

2. Gemma Church, "How Hacking the Human Heart Could Replace Pill Popping," BBC News, December 16, 2019, https://www.bbc.com/future/article/20191216-how-hacking-the-human-heart-could-replace-pill-popping.

3. "What Is AlphaFold?," EMBL's European Bioinformatics Institute, https://www.ebi.ac.uk/training/online/courses/alphafold/an-introductory-guide-to-its-strengths-and-limitations/what-is-alphafold/.

4. "The Generations Study," The Institute of Cancer Research, https://www.icr.ac.uk/research-and-discoveries/other-strategic-partner-centres/the-generations-study.

5. Denis Campbell, "Patient Privacy Fears as US Spy Tech Firm Palantir Wins £330m NHS Contract," *The Guardian*, November 21, 2023, https://www.theguardian.com/society/2023/nov/21/patient-privacy-fears-us-spy-tech-firm-palantir-wins-nhs-contract.

6. Paddy Iyer, "Real-Time Healthcare: Unveiling Patient Journeys with Stream Processing & Data Vault 2.0," *Medium* (blog),

February 3, 2024, https://medium.com/@bqqsqjzfy/real-time-healthcare-unveiling-patient-journeys-with-stream-processing-data-vault-2-0-b3658e524011.

7. Jordan Scott, "What Are Digital Twins and How Can They Be Used in Healthcare?," *Health Tech Magazine*, January 10, 2024, https://healthtechmagazine.net/article/2024/01/what-are-digital-twins-and-how-can-they-be-used-healthcare.

8. *FCC's David A. Bray on Sharing behind the Scenes Data Successes*, 2015, https://www.youtube.com/watch?v=8urW2NMgNGI.

9. Engineering National Academies of Sciences et al., "Health Worker Production," *Evaluation of PEPFAR's Contribution (2012–2017) to Rwanda's Human Resources for Health Program* (National Academies Press (US), 2020), https://www.ncbi.nlm.nih.gov/books/NBK558442/.

10. Adam Baker et al., "A Comparison of Artificial Intelligence and Human Doctors for the Purpose of Triage and Diagnosis," *Frontiers in Artificial Intelligence* 3 (November 30, 2020): 543405, https://doi.org/10.3389/frai.2020.543405.

11. Katia Riddle, "The (Artificial Intelligence) Therapist Can See You Now," NPR, April 7, 2025, https://www.npr.org/sections/shots-health-news/2025/04/07/nx-s1-5351312/artificial-intelligence-mental-health-therapy.

12. Robbie Gonzalez, "Virtual Therapists Help Veterans Open Up About PTSD," *Wired*, October 17, 2017, https://www.wired.com/story/virtual-therapists-help-veterans-open-up-about-ptsd/.

13. Office of the Attorney General, Departments of Justice California, "California Attorney General's Legal Advisory on the Application of Existing California Law to Artificial Intelligence in Healthcare," May 13, 2024, chrome-extension://efaidnbmnnnibpcajpcgl clefindmkaj/https://oag.ca.gov/system/files/attachments/press-docs/Final%20Legal%20Advisory%20-%20Application%20of%20 Existing%20CA%20Laws%20to%20Artificial%20Intelligence%20 in%20Healthcare.pdf.

14. Paul Kinahan, "FDA Publishes List of AI-Enabled Medical Devices," UW Medicine - Department of Radiology, https://rad.washington.edu/news/fda-publishes-list-of-ai-enabled-medical-devices/.

15. "Affective Computing," *Wikipedia*, March 6, 2025, https://en.wikipedia.org/w/index.php?title=Affective_computing&oldid=1279072342.

16. Satyajit Sinha, "State of IoT 2024: Number of Connected IoT Devices Growing 13% to 18.8 Billion Globally," IoT Analytics, September 3, 2024, https://iot-analytics.com/number-connected-iot-devices/.

17. "Highbridge Park News - NYC PARKS LAUNCHES SMART BENCH PILOT PROGRAM:," New York City Department of Parks and Recreation, May 24, 2016, https://www.nycgovparks.org/parks/highbridge-park/pressrelease/21372.

18. Joe Svetlik, "Is Your TV Spying on You? What Your TV Knows About You and How to Stop It Snooping," What Hi-Fi, August 16, 2024, https://www.whathifi.com/features/is-your-tv-tracking-you-what-your-tv-knows-about-you-and-how-to-stop-it-snooping.

19. Tanweer Alam, "A Reliable Communication Framework and Its Use in Internet of Things (IoT)," *International Journal of Scientific Research in Computer Science, Engineering and Information Technology* 3, no. 5 (May 30, 2018): 450–56.

20. Andy Greenberg, "Hackers Remotely Kill a Jeep on the Highway – With Me in It," *Wired*, https://www.wired.com/2015/07/hackers-remotely-kill-jeep-highway/.

21. Tehseen Mazhar et al., "Analysis of IoT Security Challenges and Its Solutions Using Artificial Intelligence," *Brain Sciences* 13, no. 4 (April 19, 2023): 683, https://doi.org/10.3390/brainsci13040683.

22. Brian X. Chen, "Ray-Ban Meta Glasses Raise Concerns About Privacy Rights, Consent and Increased Surveillance," Business & Human Rights Resource Centre, December 13, 2023, https://www.business-humanrights.org/en/latest-news/ray-ban-meta-glasses-raise-concerns-about-privacy-rights-consent-and-increased-surveillance/.

23. Lindsey Choo, "Two Harvard Students Use Meta Ray-Bans and AI to Get Personal Information," *Forbes*, October 4, 2024, https://www.forbes.com/sites/lindseychoo/2024/10/04/meta-ray-bans-ai-privacy-surveillance/.

24. Kevin Poireault, "US Legislators Demand Transparency in Apple's UK Backdoor Court Fight," *Infosecurity Magazine*, March 17, 2025, https://www.infosecurity-magazine.com/news/us-legislators-transparency-apple/; Leander Kahney, "The FBI Wanted a Backdoor to the iPhone. Tim Cook Said No," *Wired*, April 16, 2019, https://www.wired.com/story/the-time-tim-cook-stood-his-ground-against-fbi/.

25. Johana Bhuiyan, "Amazon Ring Says US Police Will Now Need Warrant to Access User Footage," *The Guardian*, January 24, 2024, https://www.theguardian.com/technology/2024/jan/24/police-warrant-amazon-ring-footage.

26. Alexander Butler, "Murderer Jailed for Life After Alexa Recordings Bring Him to Justice," *Mail Online*, March 24, 2023, https://www.dailymail.co.uk/news/article-11899217/Murderer-jailed-life-voice-recordings-Amazon-device-helped-bring-justice.html.

27. Minyvonne Burke, "Amazon's Alexa May Have Witnessed Alleged Florida Murder, Authorities Say," NBC News, November 2, 2019, https://www.nbcnews.com/news/us-news/amazon-s-alexa-may-have-witnessed-alleged-florida-murder-authorities-n1075621.

28. Anthony Robledo, "Amazon Is Removing an Echo Privacy Setting That Keeps Alexa Recordings from the Company," *USA TODAY*, March 17, 2025, https://www.usatoday.com/story/tech/2025/03/17/amazon-echo-alexa-reporting-privacy/82503576007/.

29. Aparna R. Gullapalli et al., "Quantifying the Psychopathic Stare: Automated Assessment of Head Motion Is Related to Antisocial Traits in Forensic Interviews," *Journal of Research in Personality* 92 (June 1, 2021): 104093, https://doi.org/10.1016/j.jrp.2021.104093.

30. Jeff Larson et al., "How We Analyzed the COMPAS Recidivism Algorithm," ProPublica, May 23, 2016, https://www.propublica.org/article/how-we-analyzed-the-compas-recidivism-algorithm.

31. Chris Baraniuk, "Why We Place Too Much Trust in Machines," BBC News, October 20, 2021, https://www.bbc.com/future/article/20211019-why-we-place-too-much-trust-in-machines.

32. Courtney Hinkle, "The End of Anonymity: How Facial Recognition Technology Will Worsen Online Harassment," *The Georgetown Journal of Gender and the Law* XXI, no. III (2020), https://www.law.georgetown.edu/gender-journal/online/volume-xxi-

online/the-end-of-anonymity-how-facial-recognition-technol
ogy-will-worsen-online-harassment/.

33. "The French SA Fines Clearview AI EUR 20 Million," European Data Protection Board, October 20, 2022, https://www .edpb.europa.eu/news/national-news/2022/french-sa-fines-clearview-ai-eur-20-million_en.

34. Drew Donnelly, "China Social Credit System Explained - How It Works [2025]," Horizons, February 11, 2024, https://joinhori zons.com/china-social-credit-system-explained/.

35. Marcella Siqueira Cassiano, "China's Hukou Platform: Windows into the Family," *Surveillance and Society* 17, no. 1/2 (March 31, 2019): 232–39, https://doi.org/10.24908/ss.v17i1/2.13125.

36. Thor Benson, "The Danger of License Plate Readers in Post-Roe America," *Wired*, July 7, 2022, https://www.wired.com/story/license-plate-reader-alpr-surveillance-abortion/.

37. Kristen Finklea, "Law Enforcement and Technology: Use of Automated License Plate Readers," Congressional Research Service, August 19, 2024, https://www.congress.gov/crs-prod uct/R48160.

38. "The Medici Archive Project," https://www.medici.org/.

39. *U.S. Military's TEST $2M Robotic Mule That Carries Stuff Like a Beast!*, 2024, https://www.youtube.com/watch?v=KcU2VtJzKD8.

40. Ann Cao, "Shanghai District Deploys Drones to Watch for Covid-19 Rule Violators," *South China Morning Post*, August 15, 2022, https://www.scmp.com/tech/big-tech/article/3188961/shanghais-yangpu-district-deploys-drones-detect-viola tions-covid-19.

41. Hillel Italie, "John Grisham, George R. R. Martin and Other Authors Sue OpenAI for Copyright Infringement," *Los Angeles Times*, September 20, 2023, https://www.latimes.com/world-nation/story/2023-09-20/john-grisham-george-r-r-martin-and-other-authors-sue-openai-for-copyright-infringement.

42. Laura Kuenssberg, "Elton John Brands Government 'Absolute Losers' over AI Copyright Plans," BBC News, May 18, 2025, https://www.bbc.com/news/articles/c8jg0348yvxo.

43. Cindy McIntyre, "Horses, Mules Contributed to Allied War Effort," www.army.mil, July 13, 2017, https://www.army.mil/article/190788/horses_mules_contributed_to_allied_war_effort.

44. "War Horse Facts," Brooke, https://www.thebrooke.org/get-involved/every-horse-remembered/war-horse-facts.

45. "Ranchers Use ATVs More than Horses - Progressive Cattle," Ag Proud, https://www.agproud.com/articles/52508-ranchers-use-atvs-more-than-horses.

46. Qiaoyi Li and Brenda Goh, "Beijing Unveils Plans to Boost Driverless Vehicle Use in Capital," Reuters, December 31, 2024, https://www.reuters.com/world/china/beijing-unveils-plans-boost-driverless-vehicle-use-capital-2024-12-31/.

47. Ezra Klein, "Now Is the Time of Monsters," *New York Times*, January 12, 2025.

48. "Global North and Global South," *Wikipedia*, June 9, 2025, https://en.wikipedia.org/w/index.php?title=Global_North_and_Global_South&oldid=1294703035.

49. Margaret Atwood, *The Handmaid's Tale* (McCleland and Stewart, 1985).

50. Meta Is Building AI That Can Write Code Like a Mid-Level Engineer, According to Mark Zuckerberg By Sherin Shibu Jan 20 2025 in Entrepreneur https://www.entrepreneur.com/business-news/meta-developing-ai-engineers-to-write-code-instead-of-humans/485806.

51. Riley Griffin, "Meta to Cut Roughly 5% of Staff, Targeting Lowest Performers," Bloomberg.Com, January 14, 2025, https://www.bloomberg.com/news/articles/2025-01-14/meta-is-planning-to-cut-5-of-lowest-performers-memo-shows.

52. Sherin Shibu, "Google Recruits AI to Write 25% of Its Code: Earnings Call," *Entrepreneur*, October 30, 2024, https://www.entrepreneur.com/business-news/google-recruits-ai-to-write-25-of-its-code-earnings-call/482167.

53. Mike Allen and Jim VandeHei, "Behind the Curtain: Ph.D.-Level AI Breakthrough Expected Very Soon," Axios, January 19, 2025, https://www.axios.com/2025/01/19/ai-superagent-openai-meta.

54. Peter Dizikes, "Daron Acemoglu: What Do We Know about the Economics of AI?," *MIT News*, December 6, 2024, https://eco

nomics.mit.edu/news/daron-acemoglu-what-do-we-know-about-economics-ai.

55. Greg Rosalsky, "What America's Top Economists Are Saying About AI and Inequality," WYPR, January 7, 2025, https://www.wypr.org/2025-01-07/what-americas-top-economists-are-saying-about-ai-and-inequality.

56. Laurence Iliff, "Tesla's Future Robot Legions: First They Learn to Drive, Then to Walk," *Automotive News*, May 20, 2025, https://www.autonews.com/technology/mobility/an-tesla-grand-robotic-plan-0520/www.autonews.com/technology/mobility/an-tesla-grand-robotic-plan-0520/.

57. "Insights," Evident, https://evidentinsights.com/insights/.

Chapter 8

1. Josh Howarth, "Generation Alpha: Statistics, Data and Trends (2025)," *Exploding Topics* (blog), June 5, 2025, https://explodingtopics.com/blog/generation-alpha-stats.

2. "Exoskeleton (Human)," *Wikipedia*, June 16, 2025, https://en.wikipedia.org/w/index.php?title=Exoskeleton_(human)&oldid = 1295825563.

3. Abhijeet Adhikari, "5 AI-Powered Assistive Technologies for People with Disabilities in US," AIM Research, June 12, 2024, https://aimresearch.co/market-industry/5-ai-powered-assistive-technologies-for-people-with-disabilities-in-us.

4. *Robot & Frank* (Dog Run Pictures, Park Pictures, TBB, 2012).

5. Hadley Gamble and Matt Clinch, "Bill Gates Says Technology Could 'Accentuate' the Gap Between the Rich and Poor," CNBC, November 15, 2017, https://www.cnbc.com/2017/11/14/bill-gates-defends-the-rise-of-the-robots.html.

6. Richard Pak and Leanne Clark-Shirley, "Designing Artificial Intelligence Technologies for Older Adults," World Economic Forum, September 3, 2021, https://www.weforum.org/publications/designing-artificial-intelligence-technologies-for-older-adults/.

Chapter 9

1. Mandy Deroche and Jacob Elkin, "How Much Do We Subsidize Cryptocurrency Mining's Electricity Use? No One Knows," Earthjustice, https://earthjustice.org/experts/mandy-deroche/how-much-do-we-subsidize-cryptocurrency-minings-electricity-use-no-one-knows.

2. "Hard and Soft Forks: A Detailed and Simplified Explanation of How Blockchains Evolve," Freeman Law, https://freemanlaw.com/hard-and-soft-forks-a-detailed-and-simplified-explanation-of-how-blockchains-evolve/.

3. Harmut Neven, "Meet Willow, Our State-of-the-Art Quantum Chip," *Google* (blog), December 9, 2024, https://blog.google/technology/research/google-willow-quantum-chip/.

4. Catherine Bolgar, "Microsoft's Majorana 1 Chip Carves New Path for Quantum Computing," Microsoft | Source, February 19, 2025, https://news.microsoft.com/source/features/innovation/microsofts-majorana-1-chip-carves-new-path-for-quantum-computing/.

5. "NIST Releases Post-Quantum Encryption Standards," *National Quantum Initiative* (blog), August 13, 2024, https://www.quantum.gov/nist-releases-post-quantum-encryption-standards/.

6. Ariel Schwartz, "In 20 Years, We Will Need a Second Earth," Fast Company, October 13, 2010, https://www.fastcompany.com/1694750/20-years-we-will-need-second-earth/.

7. Scott Rosenberg, "Why Tariffs Are Kryptonite to the AI Business," Axios, April 8, 2025, https://www.axios.com/2025/04/08/trump-tariffs-ai-investment-infrastructure-jobs.

8. Kay Firth-Butterfield, "Ensuring Impactful (And Safe) Innovation," The Innovator, November 15, 2024, https://theinnovator.news/ensuring-impactful-and-safe-innovation/.

9. Marc Andreessen, "The Techno-Optimist Manifesto," *Andreessen Horowitz* (blog), October 16, 2023, https://a16z.com/the-techno-optimist-manifesto/.

10. Elizabeth Spiers, "A Tech Overlord's Horrifying, Silly Vision for Who Should Rule the World," *New York Times*, October 28, 2023,

https://www.nytimes.com/2023/10/28/opinion/marc-andrees sen-manifesto-techno-optimism.html.

11. Timothy B. Lee, "Why Tech Billionaire Peter Thiel Just Endorsed Donald Trump," VOX, July 21, 2016, https://www.vox.com/ 2016/7/21/12216620/peter-thiel-donald-trump.

12. Antonio Pequeño IV, "What We Know About J. D. Vance's Relationship with Billionaire GOP Donor Peter Thiel," *Forbes*, July 16, 2024, https://www.forbes.com/sites/antoniopeque noiv/2024/07/16/jd-vance-and-peter-thiel-what-to-know-about-the-relationship-between-trumps-vp-pick-and-the-billionaire/.

13. "Understanding TESCREAL with Dr. Timnit Gebru and Émile Torres," Dave Roy Presents, https://podcasts.apple.com/nl/pod cast/understanding-tescreal-with-dr-timnit-gebru-and/ id1610914569?i=1000617036014.

14. Christopher G. Thomas, "Project Prakash Enlightens Our Understanding of Vision," NIH | National Eye Institute, January 2, 2025, https://www.nei.nih.gov/about/goals-and-accomplishments/ nei-research-initiatives/international-vision-research/project-prakash-enlightens-our-understanding-vision.

15. David Marchese, "Curtis Yarvin Says Democracy Is Done. Powerful Conservatives Are Listening," *New York Times*, January 18, 2025, https://www.nytimes.com/2025/01/18/magazine/cur tis-yarvin-interview.html.

16. Spiers, "A Tech Overlord's Horrifying, Silly Vision for Who Should Rule the World."

17. Thomas H. Davenport and Randy Bean, "Five Trends in AI and Data Science for 2025," *MIT Sloan Management Review*, January 8, 2025, https://sloanreview.mit.edu/article/five-trends-in-ai-and-data-science-for-2025/.

18. Thomas H. Davenport and Randy Bean, "Five Trends in AI and Data Science for 2025," *MIT Sloan Management Review*, January 8, 2025, https://sloanreview.mit.edu/article/five-trends-in-ai-and-data-science-for-2025/.

19. Léo Boisvert et al., "WorkArena++: Towards Compositional Plannin and Reasoning-Based Common Knowledge Work Tasks," February 5, 2025, arXiv:2407.05291v2.

20. Scott Rosenberg, "Tech Titans Want to Convince CEOs That AI Agents Will Staff Their Companies," Axios, January 23, 2025, https://www.axios.com/2025/01/23/ai-coworkers-agents-beni off-altman.

21. Jonny Hart, "AI Won't Take Your Job; It Will Make You Better at It," Temple Now, January 16, 2025, https://news.temple.edu/ news/2025-01-16/ai-won-t-take-your-job-it-will-make-you- better-it.

22. "Claude Sonnet 3.7 (Often) Knows When It's in Alignment Evaluations," *Apollo Research* (blog), https://www.apollore search.ai/blog/claude-sonnet-37-often-knows-when-its-in-align ment-evaluations.

23. Nick Bostrom, "Ethical Issues in Advanced Artificial Intelligence," https://nickbostrom.com/ethics/ai.

24. Jeff Dutton, "The AI Paperclip Problem Explained," *Medium* (blog), May 16, 2023, https://medium.com/@jeffreydutton/the- ai-paperclip-problem-explained-233e7e57e4e3.

25. "AI - Stuart Russell on Radio Davos," Radio Davos, https:// www.weforum.org/podcasts/radio-davos/episodes/ai-stu art-russell/.

26. Jennifer L. Schenker, "Agentic AI: Artificial Intelligence's Next Wave," The Innovator, October 25, 2024, https://theinnova tor.news/agentic-ai-artificial-intelligences-next-wave/.

27. "Lattice," https://lattice.com.

28. Rosenberg, "Tech Titans Want to Convince CEOs That AI Agents Will Staff Their Companies."

29. Gary Marcus, "AI Coding Fantasy Meets Pac-Man," *Marcus on AI* (blog), March 12, 2025, https://garymarcus.substack.com/p/ ai-coding-fantasy-meets-pac-man.

30. "Executive Order 14110: Safe, Secure, and Trustworthy Development and Use of Artificial Intelligence," *Federal Register*, October 31, 2023, https://www.federalregister.gov/documents/ 2023/11/01/2023-24283/safe-secure-and-trustworthy-develop ment-and-use-of-artificial-intelligence.

31. "Presidential Actions: Removing Barriers to American Leadership in Artificial Intelligence," The White House, January 23, 2025, https://www.whitehouse.gov/presidential-actions/2025/ 01/removing-barriers-to-american-leadership-in-artificial- intelligence/.

32. Christine Mui, "What AGI Hype Means for Washington," POLITICO, January 8, 2025, https://www.politico.com/newsletters/digital-future-daily/2025/01/08/what-agi-hype-means-for-washington-00197154.

33. Future of Life Institute, "Pause Giant AI Experiments: An Open Letter," March 22, 2023, https://futureoflife.org/open-letter/pause-giant-ai-experiments/.

34. Ilana Golbin Blumenfeld and Maria Luciana Axente, "9 Ethical AI Principles for Organizations to Follow," World Economic Forum, June 23, 2021, https://www.weforum.org/stories/2021/06/ethical-principles-for-ai/.

35. Matthew Kosinksi, "What Is Black Box AI and How Does It Work? | IBM," October 29, 2024, https://www.ibm.com/think/topics/black-box-ai.

36. *The Terminator* (Cinema '84, Euro Film Funding, Hemdale, 1984).

37. *The Matrix* (Warner Bros., Village Roadshow Pictures, Groucho Film Partnership, 1999).

38. Jonathan Davis, "Understanding Anthropic's Golden Gate Claude," *Medium* (blog), June 27, 2024, https://medium.com/@jonnyndavis/understanding-anthropics-golden-gate-claude-150f9653bf75.

39. *The Bletchley Declaration by Countries Attending the AI Safety Summit, 1–2 November 2023*, February 13, 2025, https://www.gov.uk/government/publications/ai-safety-summit-2023-the-bletchley-declaration/the-bletchley-declaration-by-countries-attending-the-ai-safety-summit-1-2-november-2023.

40. Dan Milmo and Eleni Courea, "US and UK Refuse to Sign Paris Summit Declaration on 'Inclusive' AI," *The Guardian*, February 11, 2025, https://www.theguardian.com/technology/2025/feb/11/us-uk-paris-ai-summit-artificial-intelligence-declaration.

41. Ezra Klein, "Now Is the Time of Monsters," *New York Times*, January 12, 2025, https://www.nytimes.com/2025/01/12/opinion/ai-climate-change-low-birth-rates.html.

42. Chris Marr, "Trump's Disparate Impact Blow Makes AI Bias Claims Even Tougher," Bloomberg Law, March 26, 2025, https://

news.bloomberglaw.com/daily-labor-report/trumps-disparate-impact-blow-makes-ai-bias-claims-even-tougher.

43. "FTC Proposes New Protections to Combat AI Impersonation of Individuals," Federal Trade Commission, February 15, 2024, https://www.ftc.gov/news-events/news/press-releases/2024/02/ftc-proposes-new-protections-combat-ai-impersonation-individuals.

44. Elham Tabassi, "*Artificial Intelligence Risk Management Framework (AI RMF 1.0),*" National Institute of Standards and Technology (U.S.), January 26, 2023, https://doi.org/10.6028/NIST.AI.100-1.

45. Blake Brittain, "Pulitzer-Winning Authors Join OpenAI, Microsoft Copyright Lawsuit," Reuters, December 20, 2023, https://www.reuters.com/legal/pulitzer-winning-authors-join-openai-microsoft-copyright-lawsuit-2023-12-20/.

46. "State Resource List: Artificial Intelligence," National Governors Association, January 2024, https://www.nga.org/wp-content/uploads/2024/01/State-Resource-List-on-AI.pdf.

47. "The One Big Beautiful Bill Section by Section," United States House Committee on Ways & Means, https://waysandmeans.house.gov/wp-content/uploads/2025/05/The-One-Big-Beautiful-Bill-Section-by-Section.pdf.

48. Marisa Garcia, "What Air Canada Lost in 'Remarkable' Lying AI Chatbot Case," *Forbes*, February 19, 2024, https://www.forbes.com/sites/marisagarcia/2024/02/19/what-air-canada-lost-in-remarkable-lying-ai-chatbot-case/.

49. "Responsible AI Principles and Approach," Microsoft AI, https://www.microsoft.com/en-us/ai/principles-and-approach; Sundar Pichai, "AI at Google: Our Principles," *Google* (blog), June 7, 2018, https://blog.google/technology/ai/ai-principles/.

50. "Reflections," *Sam Altman* (blog), January 5, 2025, https://blog.samaltman.com/reflections.

51. "Armilla: Insurance for AI Risks," https://www.armilla.ai/.

52. "OpenAI O1 System Card" OpenAI, https://openai.com/index/openai-o1-system-card/.

53. Tharin Pillay, "New Tests Reveal AI's Capacity for Deception," *TIME*, December 15, 2024, https://time.com/7202312/new-tests-reveal-ai-capacity-for-deception/; Jeremy Kahn, "Top AI Labs

Aren't Doing Enough to Ensure AI Is Safe, a Flurry of Recent Datapoints Suggest," *Fortune*, December 17, 2024, https://fortune.com/2024/12/17/openai-o1-deception-unsafe-safety-testing-future-of-life-institute-grades/.

54. "FLI AI Safety Index 2024," Future of Life Institute, December 11, 2024, https://futureoflife.org/wp-content/uploads/2024/12/AI-Safety-Index-2024-Full-Report-27-May-25.pdf.

55. Hugh Lofting, *The Story of Doctor Dolittle: Being the History of His Peculiar Life at Home and Astonishing Adventures in Foreign Parts: Never Before Printed* (Frederick A. Stokes, 1920).

56. Briana Pobiner, "Evidence for Meat-Eating by Early Humans," *Nature Education Knowledge* 4, no. 6 (2013): 1, https://www.nature.com/scitable/knowledge/library/evidence-for-meat-eating-by-early-humans-103874273/.

57. Helen Pilcher, "Elephants Have Names for Each Other, Just like Humans, Study Finds," *Discover Wildlife*, June 14, 2024, https://www.discoverwildlife.com/animal-facts/elephants-have-names-for-each-other.

58. Andrea Rosati, "Guiding Principles of AI: Application in Animal Husbandry and Other Considerations," *Animal Frontiers: The Review Magazine of Animal Agriculture* 14, no. 6 (January 4, 2025): 3–10, https://doi.org/10.1093/af/vfae045.

59. "Earth Species Project," https://www.earthspecies.org/.

60. Gary Marcus, "GPT-5: Overdue, Overhyped and Underwhelming. And That's Not the Worst of It," August 10, 2025, https://garymarcus.substack.com/p/gpt-5-overdue-overhyped-and-underwhelming?publication_id=888615&post_id=170534403&isFreemail=true&r=2s00v&triedRedirect=true.

61. https://en.wikipedia.org/wiki/Transcendence_(2014_film).

62. https://www.theosthinktank.co.uk/research/2024/02/15/ai-and-the-afterlife-from-digital-mourning-to-mind-uploading.

63. Cade Metz, "Invasion of the Home Robot," *New York Times*, April 4, 2025, https://www.nytimes.com/2025/04/04/technology/humanoid-robots-1x.html.

64. Ibid.

65. Claudio Novelli et al., "AI as Legal Persons: Past, Patterns, and Prospects," *SSRN Scholarly Paper*, November 24, 2024, https://doi.org/10.2139/ssrn.5032265.

Chapter 10

1. *The Social Dilemma* (Exposure Labs, Argent Pictures, The Space Program, 2020).

2. "The Project Gutenberg eBook of *The Federalist Papers*, by Alexander Hamilton and John Jay and James Madison," http://mirror.uwaterloo.ca/gutenberg/1/18/18-h/18-h.htm.

Appendix

1. Jena Hilliard, "Social Media Addiction," Addiction Center, March 14, 2025, https://www.addictioncenter.com/behavioral-addictions/social-media-addiction/.

2. "Artificial Intelligence and Combatting Online Child Sexual Exploitation and Abuse," Homeland Security Investigation, https://www.dhs.gov/sites/default/files/2024-09/24_0920_k2p_genai-bulletin.pdf; "The Growing Concerns of Generative AI and Child Sexual Exploitation," *National Center for Missing & Exploited Children* (blog), December 13, 2024, http://www.ncmec.org/content/ncmec/en/blog/2024/the-growing-concerns-of-generative-ai-and-child-sexual-exploitation.html; Bracket Foundation, UNICRI Centre for AI and Robotics, and Value for Good GmbH, "Generative AI: A New Threat for Online Child Sexual Exploitation and Abuse," Bracket Foundation, 2024, https://unicri.org/sites/default/files/2024-09/Generative-AI-New-Threat-Online-Child-Abuse.pdf.

3. John Schultz, "HPE Teams Up to Power Cybersecurity for Modern-Day Slavery Survivors," May 14, 2024, https://www.hpe.com/us/en/newsroom/blog-post/2024/05/powering-cybersecurity-for-modern-day-slavery-survivors.html; "Technology and Human Trafficking: Fighting the Good Fight," *Thomson Reuters Law* (blog), January 2, 2025, https://legal.thomsonreuters.com/blog/technology-and-human-trafficking/.

4. "Recruitment of Child Soldiers Is on the Rise, Despite Global Commitments," *UN News*, December 31, 2024, https://news.un.org/en/story/2024/12/1158661.

Acknowledgments

With many thanks to those who contributed thought leadership to the book, particularly Cecile Aptel, Gaby Salazar, and Ann Ewasechko.

Thanks to those who read and commented on the book, which made it much better, including my husband, Walter Burrough, PhD, who did preliminary edits.

And thanks to all the team at Wiley whose faith in me and support has been second to none.

About the Author

Kay Firth-Butterfield is the CEO of Good Tech Advisory. She is a *TIME Magazine* 100 Impact Awardee 2024 for her work in AI, Who's Who Emerging Innovator 2024, Who's Who in America, and Forbes 50 over 50 awardee 2024. She was the inaugural and former head of Artificial Intelligence & Quantum and member of the executive committee at the World Economic Forum and, in 2014, became the world's first chief AI ethics officer. She is one of the foremost experts in the world on the governance of AI and how to integrate AI into business and government to create effective transformation. Kay is a barrister, former judge, professor, technologist, and entrepreneur who has an abiding interest in how humanity can benefit from new technologies, especially AI. Kay is also an executive advisor to the Cantellus Group, visiting scholar at the University of Notre Dame, joint head of the AI Group at Doughty Street Chambers, London, and Fellow at New America in the Future of Work and Innovation team. She has authored books on human rights, AI, and modern slavery. Kay was vice-chair of the ground-breaking Institute of Electrical and Electronics Engineers (IEEE) Global Initiative for Ethical considerations in artificial intelligence and autonomous systems and a member of the group that met at Asilomar to create the Asilomar AI Ethical Principles. She sits on the Polaris Council for the Government Accountability Office (USA); the advisory board for UNESCO International Research Centre on AI, Swiss Re, and Fathom; the AI and Philanthropy Steering Committee of the Partnership on AI; the UKRAI TAS Hub; and the board of Earth Species Project. She is also on the World Economic Forum's Global Future Council on AI and the AI Governance Alliance & Digital Healthcare Transformation Initiative. She is an ambassador for the Global Trust Challenge co-created by the Organisation for Economic Co-operation and Development and IEEE. Kay has advanced degrees in law and international relations and regularly speaks to international audiences addressing many aspects of the beneficial and challenging technical, economic, legal, geopolitical, and social changes arising from the use of AI. Kay is a Master of the Inner Temple, London, served on the Lord Chief Justice's Advisory Panel on AI and Law and is a Liveryman of the City of London. She has been consistently recognized as a leading woman in AI since 2018 and was featured in the *New York Times* as one of 10 Women Changing the Landscape of Leadership.

Index